CYBER RISK

Research in Sociotechnical Cyber Security

Series Editors: **Genevieve Liveley**, University of Bristol and **Lizzie Coles-Kemp**, Royal Holloway, University of London

---

This series provides a unique platform for interdisciplinary work that examines the intersection of cyber security technologies and societies. Created in partnership with the national Research Institute for Sociotechnical Cyber Security, the series features work produced under the auspices of the Institute as well as specially commissioned books that will be of interest to the cyber security community.

Also available in the series

Understanding Digital Responsibilities
by **Lizzie Coles-Kemp** and **Mark Burdon**

Find out more at:

bristoluniversitypress.co.uk/
research-in-sociotechnical-cyber-security

TIM STEVENS

CYBER RISK

Managing Uncertainty in a Digital World

BRISTOL
UNIVERSITY
PRESS

First published in Great Britain in 2026 by

Bristol University Press
University of Bristol
1–9 Old Park Hill
Bristol
BS2 8BB
UK
t: +44 (0)117 374 6645
e: bup-info@bristol.ac.uk

Details of international sales and distribution partners are available at bristoluniversitypress.co.uk

DOI: 10.51952/9781529256048

British Library Cataloguing in Publication Data
A catalogue record for this book is available from the British Library

ISBN 978-1-5292-5602-4 hardcover
ISBN 978-1-5292-5603-1 ePub
ISBN 978-1-5292-5604-8 OA PDF

Cover design: Nicky Boroweic Design
Front cover image: Unsplash/Kaiyu Wu
Bristol University Press uses environmentally responsible print partners.
Printed and bound in Great Britain by CPI Group (UK) Ltd, Croydon, CR0 4YY

Bristol University Press' authorised representative in the European Union is: Easy Access System Europe, Mustamäe tee 50, 10621 Tallinn, Estonia,
Email: gpsr.requests@easproject.com

FSC
www.fsc.org
MIX
Paper | Supporting responsible forestry
FSC® C013604

Contents

Series Editors' Preface

Genevieve Liveley and Lizzie Coles-Kemp

The aim of this series of short books is to provide the first dedicated platform from which to showcase leading-edge interdisciplinary research focused on *sociotechnical* cyber security. The series features research produced under the auspices of the UK's Research Institute for Sociotechnical Cyber Security as well as specially commissioned works of interest to the national and international security communities. The books published as part of this series recognize that technology-centric approaches to cyber security aren't enough on their own. They introduce the latest research and new ideas for transforming security thinking so that *people* are empowered to be the first and the last line of defence. They address the bigger sociotechnical approaches to cyber security – the cultural, psychological and behavioural influences that make people act as they do in real world security scenarios.

In this volume, *Cyber Risk: Managing Uncertainty in a Digital World*, Tim Stevens investigates the imbrication of powers in cyber risk management (CRM) in and across its various forms. He presents a fresh theoretical and empirical approach to cyber risk through the lens of governmentality, arguing that both populations and risk are governed through complex combinations of political strategies, institutions, techniques and technologies. If effective CRM requires the agency and energy of powers beyond that of governments alone, the book asks, what are the political effects of CRM?

About the Author

Tim Stevens is Reader in International Security at the Department of War Studies, King's College London, and co-director of its Cyber Security Research Group. His research interests include cyber security politics, cyber risk, cyberwarfare and the global politics of technology. He has published widely on these topics, including *Cyberspace and the State* (Routledge, 2011), *Cyber Security and the Politics of Time* (Cambridge University Press, 2016), *What Is Cybersecurity For?* (Bristol University Press, 2023) and *Research Handbook on Cyberwarfare* (Edward Elgar, 2024). He holds visiting appointments at Conservatoire national des arts et métiers (Paris), Universidad Nacional de Educación a Distancia (Madrid) and the Research Institute for Sociotechnical Cyber Security (UK).

Acknowledgements

This book is the product of conversations with friends and colleagues too numerous to name. I thank them all, as I do the staff and students of the Department of War Studies, King's College London (KCL) and the KCL Cyber Security Research Group. Early research support was provided by Joe Jarnecki, and conversations with Sarah Backman, Karsten Friis and Erik Reichborn-Kjennerud were instrumental in the book's formation. I am grateful to the organizers and participants of events in the UK, Greece, Ireland, Netherlands, Spain and online where elements of this book were presented and discussed. Thank you to the series editors, Genevieve Liveley and Lizzie Coles-Kemp, for the opportunity to collaborate with them on this book, and to the anonymous reviewers for their comments and suggestions. Sincere thanks to Paul Stevens and the Bristol University Press team for their outstanding professionalism.

Brandon Valeriano, my late friend and fellow traveller, once wrote that cyber security researchers should embrace 'feeling stupid from time to time' to expand knowledge in our field. I have experienced the former frequently, although cannot claim to have achieved the latter. This book is dedicated to him.

ONE

Introduction

In January 2025, the World Economic Forum (WEF), the global political-economic advocacy organization, published the 20th edition of its 'Global risks report'. Unsurprisingly, extreme weather, armed conflict, misinformation and disinformation and societal polarization figured prominently in its severity ranking of global risks over a two-year period, joined by ecosystem collapse, artificial intelligence (AI) and natural resource shortages over the next decade (WEF, 2025a: 12). Nestled at number five among the most disturbing short-term global trends – and ninth over the ten-year period – were 'cyberespionage and warfare', where 'cyber' indicated the use of computers as tools and targets of transnational malfeasance and mischief. Lower down, there were references to 'crime and illicit economic activity', 'online harms', 'censorship and surveillance' and disruptions to 'critical infrastructure' and 'systemically important supply chains'. As the report noted, these issues were closely interlinked with one another and with other risks identified as significant in the near- and mid-term. If we take these categories of risk as functions of digital transformation and the embeddedness of digital computing technologies ('cyber') across myriad socioeconomic and political sectors, small wonder that 'elevated cyber risk perceptions' were a feature of the WEF's global outlook (WEF, 2025a: 17).

Such documents indicate important facets of 'cyber risk'. The first is its diversity. There is no widely acknowledged definition of cyber risk, so it is explained differently in every context in which it is used, including within and between expert communities (Cains et al, 2022). The UK-based Institute of Risk Management (nd) defines it as 'any risk of financial loss, disruption or damage to the reputation of an organisation from some sort of failure of its information technology systems'. The US National Institute of Standards and Technology, a leading developer of cyber risk management (CRM) frameworks, suggests cyber risk is 'the risk of depending on cyber resources (i.e., the risk of depending on a system or system elements that exist in or intermittently have a presence in cyberspace)' (Ross et al, 2021: 62). The sources of risk are different in each case – IT failure in the first, IT dependency in the second – yet they both speak to the potential destabilization or exploitation of dependence on IT systems, such that accidental or purposive actions can lead to loss, harm, damage or disruption to anything or anyone that relies upon the proper functioning of those systems. In WEF terms, cyber risk emerges in a wide range of forms: from cybercrime, cyberespionage and cyberwarfare to disruption of supply chains and critical infrastructures and intersects with adjacent fields like online harms and the problems of misinformation and disinformation. CRM has emerged as a way of managing these cyber risks, to which all entities, from the individual to the firm, states and the international system, are exposed.

The second point pertains to a key term in the WEF's methodology: perceptions. The WEF report is based on its annual 'Global risks perception survey'. For the 2025 edition, the WEF 'brought together the collective intelligence of over 900 global leaders across academia, business, government, international organizations and civil society'. It also drew on its 'Executive opinion survey' of over 11,000 business leaders in 121 countries, plus over 100 'thematic experts' in the global risk analysis community (WEF, 2025a: 5). The 'Global risks report', like other analyses of its type, captures not an objective condition of global risk but subjective impressions of its nature, character and import. This does not mean that risks are misidentified or that the judgements of the WEF's interlocutors are somehow incorrect or invalid, only that any analysis can only capture perceptions of risks (Slovic et al, 2004). Like 'security', which is conventionally framed as a response to the perception of 'threats' (Welch, 2022: 25–31), risk management can only operate if issues are perceived as risks. Dangers and hazards may have an objective reality external to the observer, but they must be perceived – or 'constructed' in social scientific terms – as threats and risks in order to be countered or managed (Power, 2007). As François Ewald (1991: 199) argues: 'Nothing is a risk in itself; there is no risk in reality. On the other hand, anything can be a risk; it all depends on how one analyzes the danger, considers the event.' Identifying something as a risk (or not) is a process of cognition, of decision and deliberation and, sometimes, emotion, as much as it is of dispassionate or scientifically disinterested observation. This has important implications for thinking about any type of risk, cyber risk included, not least that the significance ascribed to risks is elastic and subject to contestation.

This leads to a third remark. The head of the UK's National Cyber Security Centre (NCSC), Richard Horne, used his inaugural public speech on cyber risk and resilience in December 2024 to warn that 'we believe the severity of the risk facing the UK is being widely underestimated' (NCSC, 2024b). Despite many years of the UK putting CRM at the heart of its national cyber posture and policy, Horne was making a political statement: more needed to be done to bridge the gap between the actions taken to address the cyber risk to the UK and the perception of that risk. Since his speech, the major outages caused by cyber incidents affecting UK household names like Jaguar Land Rover, Marks & Spencer and the Co-op have sharpened this critique still further (Martin, 2025). Like the mobilization of cyber threats to facilitate cyber security, an inherently political dynamic (Dunn Cavelty, 2013), the perception of cyber risk drives CRM, although that perception is not normally considered an element of cyber risk itself

(Aven, 2012). Cyber risk is political because it enables forms of action that are contested, such as policy and resource allocation, and have meaningful effects on people's lives. This includes the development of a political economy of global CRM, as well as the more subtle modulations of professional and personal activities and opportunities. The main purpose of this book is to analyse the political aspects of CRM by problematizing cyber risk as a set of orientations and practices that are not just responses to the complexities of digital transformation but shape the world in profound and unexpected ways. The following section describes the three main interventions the book seeks to make.

Problematizing cyber risk

The first objective is to rescue the analysis of cyber risk from the gravity well of organizational and technical CRM. The existing cyber risk literature is dominated by work prescribing how to tackle cyber risk through managerial and organizational practice. For the corporate executive, public-sector administrator or entrepreneur, this work is full of good advice and proposes lines of effort that, if followed correctly, should reduce their organizations' exposure to cyber risk and improve their security and resilience. These authors do not declare this task is easy, and each describes how leaders must realign their processes and practices with established principles of risk management. They impress urgency upon their readers and present them with 'how to' programmes for their respective organizational settings. Many are non-technical guides aimed at management professionals (for example Evans, 2019; Falco and Rosenbach, 2022); still more are designed for technical professionals operating at the 'pointy end' of information security. It does not diminish this substantial and sophisticated body of work to suggest that the priorities of this book are markedly different.

Most analyses of cyber risk are what Robert Cox (1981: 128–9) famously called 'problem-solving', rather than 'critical', approaches to issues of power and security. Inasmuch as this body of work expresses specific theoretical frameworks at all, it reproduces the very assumptions left unexamined as the basis for these enquiries, a circular dynamic that says more about what things are than how they became so. In these accounts, cyber risk is an objective and urgent fact, an exogeneous or systemic presence to be identified, quantified and reduced through the diverse means and ways of CRM. It tells us little about how cyber issues become understood as risks, nor of the implications of this risk framing for power, security and governance. The managerial and mainstream security literatures treat cyber risk as an ontological given, not as something that has to be constructed in social contexts *as risk*. Enquiries into the sources of cyber risk, for instance, find answers in the intentionality of adversaries and criminals, and in the imperfections of systems design and use, but rarely question the epistemic origins of cyber risk as a social construction. As Deuchars (2004: 140) argues, explicitly referencing Coxian problem solvers, risk managers see risk as a problem to be solved but 'do not entertain the notion that the basis of their approach may be at fault ... such thoughts are not within the boundaries of the thinkable'. Without a shift from 'problems' of cyber risk to the 'problematization' of cyber risk, it is neither possible to account for the social factors leading to the construction of cyber risk, nor of its discursive-material effects on society.

Rather than take cyber risk as a problem to be solved, this book problematizes cyber risk as an object of enquiry, asking how cyber risk and its attendant practices can be understood as fields of social action and with what consequences. If the management literature takes the existence of cyber risk as the foundation of analysis, the starting point of this book is to enquire after the construction of cyber risk as a form of risk in the first place, the rationalities and logics that enable this process and the political effects of cyber risk and its management. Although the CRM literature is clear that identifying risks is the necessary first step in addressing cyber risk, it says little about why an issue is identified as a risk. This is because its claims to objectivity and quantification do not permit recognition of the social context of risk construction. Cyber risk does not exist pristine in the world but must be socially constructed in order to be instrumentalized, managed and governed. This argument contrasts vividly with the overwhelming assumption in risk analysis that cyber risks exist prior to their identification as risks. The causality of risk emphasized in this book is the inverse of the common perspective in which 'cyber risk manifests itself in an interconnected world' (US Cybersecurity and Infrastructure Agency, nd). It is risk practices and practitioners that manifest risk, not the other way around. Problems must be understood as risks before they can be treated as risks, which does not deny the existence of problems but affirms the perception and social construction of risks as risks and, consequently, the ability to interact with them as 'governable entities' (Dean, 1998).

The second contribution of the book is to widen the aperture of what we recognize as the operative agents and locales of CRM. If we associate cyber risk solely with the formalized practices of enterprise CRM, for instance, we restrict ourselves to the analysis of a specific set of risk professionals and practitioners operating at the level of the organization or firm. Important though this would be (see Da Silva and Jensen, 2022), the ways in which CRM is framed, designed and implemented are not restricted to the technical practices of information security or classical risk management. It is intricately interwoven with wider societal worries about safety and security and is the object of national and international public policy. This implicates a much wider range of actors and organizations and extends the operations of CRM across numerous socioeconomic and political fields, all of which are potentially important sites of scrutiny. Integral to this mosaic is a perspectival shift from threat-based to risk-based security practices.

Cyber threats and risks co-exist in what is often called the 'cyber security ecosystem'. The critical relationship between

environment. To address this, governments are increasingly 'societalizing' cyber security and cyber risk by enrolling greater numbers of non- and sub-state domestic actors into 'whole of society' security frameworks. These capitalize on latent skills and expertise but also risk constructing actors as unwilling agents of security or even war in a complex geopolitical cyber landscape.

The closing chapter, 'Conclusion', brings together the themes of the book and looks towards the future of CRM and further research. In addition to summarizing the main argument regarding the governmentality of CRM, it argues that this approach provides us with three main points of departure from the mainstream literature: shifts in perspective, the loci of agency and issues arising from preventive risk management. The role of AI in CRM is addressed, underscoring the sociotechnical nature of the cyber risk *dispositif* and the operations of power in complex assemblages of humans and nonhumans. The book concludes with final comments about the provisional character of its analysis and the hope that other researchers will investigate CRM and explore the operations of power in the governmentality of cyber risk.

TWO

Theorizing Cyber Risk

Cyber risk management (CRM) and cyber security are mutually linked and overlapping concepts that can be difficult to tell apart in practice. Each is concerned with the welfare of an entity, whether it be an organization, infrastructure, polity or population. Their practices and priorities are often deeply intertwined, and the policies that shape and constrain them frequently address them together. They are complementary rather than antagonistic, even if they are not immune from disputes over their relative importance, emphasis and resourcing. In theory, though, cyber security and cyber risk are categorically at variance, informed by differential logics of thought and behaviour. If cyber security entails the thwarting of cyber threats to an entity, cyber risk engages with the features of that entity: what vulnerabilities does it have and what harms may arise from those vulnerabilities if exploited successfully? Cyber security builds upon knowledge of known threats and attempts to secure an entity against them. CRM constructs knowledge in a more possibilistic, and even fatalistic, register: we can only provide a limited level of security, so what happens when that security fails in the future? The assumptions built into cyber security are not directly transferable to understanding the foundations of cyber risk and its management.

This chapter proceeds in three stages. The first section outlines the nature and priorities of cyber security, which principally entail the securing of an object against a range of cyber security threats. Cyber security remains a crucial element of risk management, protecting systems and patching vulnerabilities, but security can never be 100 per cent. As a result, risk has become an important overarching rubric for managing that which cannot be secured. The second section outlines how cyber risk has emerged in national and corporate policy, set against the backdrop of sociopolitical, economic and technological complexity and uncertainty. Recognition of cyber security as a 'wicked problem' has led to the adoption of risk management tools for addressing vulnerability and harm stemming from digital dependencies, assisted by the linking of cyber risk to enterprise risk, geopolitical risk and systemic risk. The third substantive section provides a theoretical account of the 'riskification' of cyber security, in which the 'problem' is constructed in the concepts and language of 'risks' rather than just 'threats'. This is a social process that enables the identification of issues as cyber risks that can then be governed through the practices of CRM. In a precursor to the main argument of the book, riskification is also a political process: it is informed by and expresses political orientations and is deployed with political effect as a form of governmentality.

From cyber security to cyber risk

Cyber security has traditionally involved the protection of data, systems and networks from threat. It channels conventional information security thinking in its aim to protect the confidentiality, integrity and availability of digital assets, the so-called 'CIA Triad', from threats to the organization from outside – and sometimes inside – its perimeter. The precise origins of this formulation are obscure, as is common in a field where informal practice frequently prefigures formal codification, which even then is often community driven rather than imposed from above. It emerged quasi-formally in the 1970s and 1980s in the United States and spread widely as a straightforward means of conceptualizing information security issues and the lines of effort needed to resolve them. 'Confidentiality' refers to the prevention of the unauthorized disclosure of information; 'integrity' to safeguarding the completeness and accuracy of information; and 'availability' requires that authorized users can access information when they need it. This approach has long been criticized for a narrow focus on technical issues at the expense of sociotechnical factors like user behaviours and organizational contexts (Dhillon and Backhouse, 2001), but it remains critical to the management of information systems and networks. Its importance is both theoretical and practical: the CIA Triad has 'not only shaped and informed our theoretical understanding of information security, but also the very practices through which security is developed and implemented in organizations' (Samonas and Coss, 2014: 22). This continues today, and when CRM discusses 'cyber security controls', these resolve to the CIA Triad in a foundational sense.

What are these 'cyber threats' to confidentiality, integrity and availability? Threats are understood in two principal ways: by threat vector and by threat actor. Threat vectors are the technical means by which an information asset or system is compromised, leading to a data breach, operational disruption or other negative outcome. These include ransomware, distributed denial of service, social engineering, malicious software (malware) and supply-chain attacks. While there is substantial diversity in the classes of entity and activity included in lists like this – for example, tools (ransomware) versus scenarios (supply-chain attack) – the security reflex is to prevent unauthorized access to networks and the harm that can result by ensuring the confidentiality, integrity and availability of digital assets. This means understanding the technical aspects of cyber threats; having situational awareness of relevant systems and networks to ascertain security vulnerabilities and the presence of, say, malware or suspicious code; and possessing the wherewithal to effect cyber security 'solutions', usually through a combination of in-house technical expertise and third-party goods and services.

This is a reasonable description of a 'classical' approach to countering cyber threats, but it is incomplete. Far from being agnostic about the source of cyber threats, public and private organizations are often keen to understand why their digital assets are being targeted and by whom. In contemporary cyber security, it helps to determine security posture and resource allocation if one understands who is seeking unauthorized access to a system and what they plan to do when they get there. This shifts the register of 'threat' from vector to actor, although the two are intimately related: one might infer an actor's intent from a defined configuration of vectors, and actors usually deploy several vectors to achieve their goals. The cyber threat intelligence (CTI) industry connects vectors and actors and enables operational and strategic decision making by identifying threats to organizations. It uses methodologies that build threat profiles from technical signatures and non-technical evidence, so that security teams can protect their assets and implement security controls in line with organizational necessity and priorities (Work, 2020; Oosthoek and Doerr, 2021).

Typologies of cyber threat actors are notoriously inconsistent in their details, but most resolve to basic categories: state (military, intelligence agencies), state proxies, cybercriminals, hacktivists, hackers and terrorists. Others prefer to distinguish between the types of activities conducted: cyberespionage (intelligence), cyberwarfare (military), cybercrime (criminals), cyberterrorism (terrorists) and various other forms of cyber-mediated mischief or political action (hackers, hacktivists) (Stevens, 2023a: 36–52). This picture is complicated by the fluidity of how actors behave and organize. Militaries conduct both intelligence and cyberwarfare operations, and intelligence agencies conduct coercive cyber operations against state and non-state targets (Gioe et al, 2020). A 'hacker' might be employed by a cybercrime group that undertakes operations on behalf of a government as its de facto proxy (Schmoldt, 2024). Hacktivists or insiders conduct operations aligned with or against a state without otherwise meaningful connection with the states in question (Munk, 2022: 208–25). Some actors engage in operations that straddle the conceptual divide between computer network operations (cyber) and information activities like disinformation and misinformation campaigns, propaganda and other forms of influence and psychological operations (Shires, 2019b; Whyte et al, 2021). In some cases, the 'Western' canonical distinctions between cyber forms of warfare, crime, espionage and subversion and information operations are unsustainable in non-Western contexts, which means these categories can be unhelpful or misleading (Dossi, 2020; Giles, 2021).

One influential approach to high-level cyber threat actors is to identify them as Advanced Persistent Threats (APTs). Developed in the US Air Force in the mid-2000s, the term migrated across government and industry to define effective, well-resourced and long-lasting threat groups, often associated with the military or intelligence agencies of hostile powers (Buzatu, 2022). APTs have political and economic motives, targeting foreign governments, strategic industries, banking, intellectual property, personal data, critical infrastructures, anything that can be leveraged for strategic or economic gain, or both. Driven by the CTI industry and broader security requirements, APT designation has become a powerful means of ascribing identity to threat groups over time. It has raised awareness of threat actors targeting specific sectors or types of organizations and enabled cyber security professionals and decision makers to adjust their defensive strategies accordingly.

It is questionable whether cyber threat typologies arranged by actors and intentions have utility to the average organization, although they serve a public policy purpose for governments seeking to frame the overall cyber threat intelligibly and to organize and justify resource expenditure and bureaucratic reorganization. At a technical level, it is often more productive to think about multiple-source CTI and what it can say about specific actor profiles – their intent, capability and opportunity – than to rely on reified high-level categories of actors, or inferences drawn from technical indicators. It is almost impossible, for instance, to discern from a single piece of software whether it has been deployed for the purpose of intelligence gathering or as a bridgehead for network exploitation (Buchanan, 2016). This complicates attribution and the conditions of decision making and response.

Cyber threats continue to occupy a vital role in how we think about and implement cyber security. Overall, though, cyber security has moved beyond a purely threat-based approach and towards understanding threats as one factor within a more holistic risk management framework. Cyber security can never be total – all sociotechnical systems have vulnerabilities, whether logical, physical or behavioural – so the spectrum of cyber security threats can never be fully countered or neutralized. As Dunn Cavelty (2008: 56) argues, if 'absolute security' is impossible, the 'logical consequence is that one has to manage the existing risks'. The following section outlines how cyber risk has entered the domain of policy as an organizing concept and principle.

Cyber risk as policy paradigm

In December 2021, the United Kingdom published the fourth iteration of its national cyber security strategy, the introduction to which heralded the 'extraordinary opportunities' of cyberspace for 'countries like the UK to pursue their national goals in new ways' (HM Government, 2021: 10). In common with most commentators on contemporary technological change, it declared that cyberspace – the notional environment created by globally networked digital technologies – 'is changing the way we live, work and communicate' and was 'transforming the critical systems we rely on in areas such as finance, energy, food distribution, healthcare and transport' (HM Government, 2021: 10). This is hard to refute. The ongoing and accelerating process of digital transformation is leaving few people unaffected by the transition from analogue to digital ways of working and living. The proliferation of online platforms and applications and our changing modes of engagement with them are facts of social life, albeit we cannot yet discern their long-term effects. But there are more profound currents too in the digitalization of critical infrastructures and the interconnection of the previously discrete – and, to the casual observer, invisible – infrastructural networks and systems upon which our daily lives depend. Data are being created in unimaginably large volumes, the growth of which is one of the few aspects of life that truly deserves the label 'exponential' (Thornton, 2024). Add to this the burgeoning Internet of Things, artificial intelligence and machine learning, nanotechnology, biotechnology, robotics, quantum computing, augmented reality and a range of other 'emerging' technologies, all of which rely on the creation and exchange of digital data, and we are, as the UK government noted, 'in the early stages of a long-term structural shift' in socioeconomic organization, driven by widespread adoption of information technologies and 'big data' (HM Government, 2021: 10).

If digital transformation presents opportunities for society and business, it also poses serious challenges. None of the technologies of cyberspace is 100 per cent secure, and the growing digital 'attack surface' can be exploited by criminals, military and intelligence agencies, hackers and adventurers for political and economic gain. Cyber security – defined by the UK as the protection of internet-connected systems, data and services from unauthorized access, harm or misuse (HM Government, 2021: 126) – has become the primary response to this situation, a suite of diverse practices that links big business with high politics and, globally, involves millions of dedicated professionals across the public and private sectors. The emphasis of cyber security is on preventing 'cyber threats' from disrupting the proper functioning of digital ('cyber') systems and networks. Most of what we associate with the everyday security of computing falls under the rubric of cyber security: anti-virus, firewalls, passwords, patching, access control, end-point protection and intrusion monitoring. There is heightened political awareness that public policy needs to galvanize and mandate better cyber security measures across enterprise and public services, while also thrashing out in international politics whether and how states should be able to exploit the social and technical vulnerabilities of cyberspace in pursuing their national ambitions. The problem of cyber security is technical, but it is also social, economic and political, a truly interdisciplinary 'wicked problem', for which a lack of lasting solutions has resulted in great complexity and instability in cyberspace and beyond (Carr and Lesniewska, 2020; Stevens, 2023a: 1–11).

As the UK government notes of digital transformation, the 'scale and speed of this change – often outpacing our social norms, laws, and democratic institutions – is also unleashing unprecedented complexity, instability and risk' (HM Government, 2021: 10). The reference to 'risk' is significant and reflects deeper dynamics in social, political and technological organization. There is a growing awareness that cyber security alone cannot be the antidote to digital disruption. The 'fortress mentality' of conventional cyber security can only go so far when, as cyber security practitioners often articulate, we have to assume that the enemy is already within our walls (Tarun, 2023). If cyber defences have already been breached and our digital assets are already under threat of theft or destruction, what then? This question has occasioned a shift to thinking about these problems not just as 'threats' but also as 'risks'. Whereas security remains mostly focused on specific internal or external threats to an entity, risk inverts the gaze to the entity itself. If a cyber threat is that which has the potential to cause harm to a system, 'cyber risk' identifies the harm that might be caused to that system. Cyber risk, in theory at least, can be modelled, quantified and, to a greater or lesser extent, embraced as a necessary condition of operating in digital space. Cyber security identifies and counters threats;

cyber risk *manages* the potential effects of a successful threat. It is, in an important sense, an acceptance that 'bad things will happen – and probably already are', a different conceptual frame to the zero-sum game of cyber security.

Cyber risk is understood to be a function of threat, vulnerability, harm and probability, although the most common formulation of cyber risk is much simpler: how likely is an event, and what impact will it have? Likelihood is the probability that a vulnerability will be exploited and is calculated – however roughly – by determining the capabilities and motivations of a threat and the effectiveness of security controls in place to counter it. This produces a coarse estimate of likelihood from low to high, or whatever graded scale a risk methodology recommends. Impact (or harm, loss, consequence) is an assessment of what will happen should a threat be successful. At its simplest, mapping likelihood against harm generates a risk assessment matrix, intended to assist decision makers in recognizing what assets are at risk, and prioritizing courses of action to secure them better. In practice, cyber risk assessment and management methodologies are usually more complex than this, and an industry has grown up to promote, advise and implement CRM as an essential facet of doing business in a digital world.

As an abstract term, 'cyber risk' is unlikely to have any 'isolated and independent meaning' outside its organizational and disciplinary context (Strupczewski, 2021: 9). At the level of practice, precisely how cyber risk and its constituent elements are understood is surprisingly granular, down to the level of individual organization, manager or analyst, even as standards and framework methodologies attempt to smooth out local variations to ensure conformity and correlation. At the conceptual level, however, the transition of cyber risk from a technical information security problem to a societal issue is connected to the discursive linkage of cyber risk with other forms of risk, principally enterprise risk, geopolitical risk and systemic risk. Holistic information assurance and

CRM frameworks have brought cyber risk into the ambit of mainstream enterprise risk management. Notwithstanding disparities in implementation and corporate culture, cyber risk is an integral part of organizational risk profiles and management postures, such that the refrain, 'cyber risk is business risk', is commonplace among the cyber security commentariat. This is not just a message for boards or IT departments: the 'notion that cyber risk is business risk must permeate throughout every layer of the enterprise' (Raissipour, 2023). Cyber risk is helping to reshape the enterprise in profound ways.

The transnational nature of digital infrastructure and the vulnerability of firms to international threat actors (criminals, APTs) mean that enterprise risk is re-tooling its understanding of risk to include the (geo)political aspects of cyber risks. A 2023 survey of Chief Information Security Officers, the executives charged with overall responsibility for cyber security, indicated that geopolitical cyber risks to the organization (33 per cent) were perceived as nearly twice as significant as criminal cyber attacks (19 per cent) (Heidrick and Struggles, 2023). This is a familiar concern of states, too, for whom the grind of cybercrime and commercial cyberespionage enervates their economies, while the persistent sub-rosa skirmishing of inter-state cyber competition demands ever more defensive and offensive resources. CRM at the global level involves multilateral and multistakeholder mechanisms and, if the United States is correct, deliberate shaping of the global information space through targeted cyber operations in contested environments (Fischerkeller et al, 2023). There is developed awareness, too, of cyber risk as systemic risk (Forscey et al, 2022; Welburn and Strong, 2022). The interconnectivity of digitally enabled systems, networks and infrastructures means, hypothetically, that failures due to error or successful attack in one system may spread to others, leading to large-scale disruptions to functionality and the development of spatially and temporally extensive crises beyond organizational and political boundaries (Backman, 2021; Ebert et al, 2023). Cyber risk is conceptually and empirically connected to other forms of significant and politically relevant risk.

Constructing cyber risk

A longstanding question for cyber security scholars is how cyber security has become a political issue. Elevating technical information security to public policy and linking the protection of critical infrastructures to national security have required significant political and material investments. Scholars disagree about whether these processes constitute a 'securitization' of cyberspace (Dunn Cavelty and Egloff, 2019). Securitization theory proposes that specific issues are singled out and labelled 'security threats' by elites, thereby shifting them from 'normal' politics into a security register that legitimizes extra-political measures to deal with them (Buzan et al, 1998; Balzacq, 2010). In contrast to classical security approaches, which explore the distribution of power and material capabilities in determining security outcomes, securitization emphasizes the performative role of discourse in constructing issues as security threats, often in existential terms. The history of cyber security has seen these moves often, most obviously in attempting to frame possible worst-case cyber security outcomes as analogous to 'existential' historical events, giving us flawed metaphors like 'cyber Pearl Harbor' and 'digital 9/11' (Stevens, 2016; Lawson, 2021). The rhetorical power of these utterances is debatable, as audiences do not blindly accept securitization moves (Balzacq, 2005; Côté, 2016). Despite the prevalence of failed securitization efforts, there is ample evidence that cyber security has been successfully moved on to national and societal security agendas in the first two decades of the 21st century (Nissenbaum, 2005; Dunn Cavelty, 2007; Whiting, 2020).

However, it is difficult to argue that cyber security has been *re*moved from democratic deliberation as an existential issue, given its prominence in public policy and the media and the significant discussion of cyber security in national and international forums. Cyber security incorporates different logics or 'grammars of securitization' that are entangled across a wide field of policy and practice (Hansen and Nissenbaum, 2009; Dunn Cavelty, 2020). Existential 'hypersecuritization' through speculative catastrophism does occur, with some effect, but attention to 'everyday practices' stresses the less spectacular securitization of citizen and consumer experiences. In addition, the epistemic authority of experts is fortified via a securitizing 'technification' of cyber security derived from its complex technical nature and origins in computer science. The interplay of these logics creates new configurations of networks and individuals that constitute the broad field of cyber security and the 'referent objects' it seeks to secure, from the state, society and nation to the economy and individuals. This warns us against unwarranted hypersecuritization and directs us towards more mundane aspects of cyber security and to the roles of non-state actors (and nonhumans) in cyber security assemblages.

Cyber security securitization is not a straightforward process of elite discourse leading to the removal of cyber security from 'normal' politics. Although hyperbolic language is still common, it has not led to the widespread adoption of emergency countermeasures associated with more 'obvious' securitizations like counterterrorism (Balzacq et al, 2016). Cyber security refers to a range of objects deemed necessary of protection and involves actors with competing preferences and objectives, only some of which follow a logic of hypersecuritization (Dunn Cavelty, 2013; Stevens, 2023a). Most cyber insecurity is not an existential threat but articulates a 'logic of subsistence': cyber security is driven by the functional needs of socioeconomic sectors, whereby 'its absence does not lead to extinction, but to dysfunction' (Ciută, 2010: 135). This dysfunction is a product not of threat actors' intent or efficacy per se but of our dependence on digital assets, whose normal operations are relied upon for everything from the most critical infrastructural services to economic productivity, communications and play. This dependence is a source of vulnerability and potential harm, which falls squarely into the paradigm of risk. Accordingly, securitization logics are accompanied by 'riskification', described as 'a social process of constructing something politically in terms of risks' (Corry, 2012: 238).

Riskification expresses three key dynamics. First, a shift away from the direct causes of harm (threats) to the 'permissive causes' of 'future possible harmful events' (Corry, 2012: 246). In contrast to 'first-order' threat politics, risk is 'a second-order security politics that directs attention to the level of conditions of danger or harm' (Corry, 2012: 246). This 'constitutive causality' argument posits that contemporary configurations of material and discursive resources are what allow future events or actions to occur. In cyber security, a risk outlook pays less heed to 'malicious actors' intent and capabilities' than to the conditions of possibility that enable cyber operations: '[I]t is the dependencies, vulnerabilities and resilience of [one's] own systems that largely define the probability and consequences of an attack' (Friis and Reichborn-Kjennerud, 2016: 37). Harm is realized not by threats but by the posture and readiness of a potential target and by the sociotechnical idiosyncrasies of cyberspace. Cyber risk is less about immediate threats and more about the 'assessment of future dangers and efforts to prevent or mitigate these' (Friis and Reichborn-Kjennerud, 2016: 37).

The second facet of riskification relates to the governance of the possibility of harm, entailing a proactive focus on as-yet unknown vulnerabilities dissimilar to security's reactive approach to known cyber threats. This prioritizes and rewards the 'technification' of cyber security, granting epistemic authority to technical experts. It also transforms cyber security into a different creature – 'this is management, not defense' (Friis and Reichborn-Kjennerud, 2016: 38). CRM operates at the organizational level and in public policy and international affairs, all aiming to govern rather than prevent cyber risk (Friis and Reichborn-Kjennerud, 2016: 38). The third component of riskification is the temporal nature of CRM, which is longer term than the greater immediacy of a threat-based approach. Reducing vulnerability is an ongoing programme involving diverse actors and directed at a wide array of problems, few of which are amenable to short-term solutions. Efforts are directed towards 'soft' measures like resilience, awareness raising, confidence building, information sharing and other precautionary actions that minimize risk profiles over time (Friis and Reichborn-Kjennerud, 2016: 39).

The riskification of 'cyber' does not imply that cyber *security* is irrelevant. Risk and security are modes of engagement with the problems of vulnerability and threat, respectively, but they co-exist and commingle in cyber security ecosystems. In the UK, for example, risk has become a central tenet of national security decision making, but it augments rather than replaces security in the policy process (Neal, 2019: 249). Like security, though, risk is deeply political. If securitization relies on discursively establishing entities as threats, riskification requires that hazards be constructed intersubjectively as risks (see Battistelli and Galantino, 2019). Cyber risks are positioned as risks to digital technologies (cyberspace) – the focus of this book – and as risks through cyberspace, such as internet-enabled terrorist radicalization and recruitment, political dissent and civil protest (Deibert and Rohozinski, 2010). Riskification can also be understood as the normalization, or desecuritization, of threat, which allows for a more expansive definition and repurposing of 'security issues' by a wider range of actors (Dunn Cavelty, 2007: 139–42; Burton and Lain, 2020; Burton and Christou, 2021). The inwards gaze of risk to the sociotechnical 'thing itself' rather than external threats may desecuritize sociotechnical problems, restoring them to democratic oversight and accountability, even if this often adopts a markedly technocratic flavour. In the reflexive terms of the 'risk society', this desecuritization is a collapse of the distinction between 'security threat' and 'security referent', as the former is generated by the latter (Chandler, 2018a: 9).

Risk management demands expansive stakeholder engagement and activity beyond the state, which implies greater transparency, information sharing and democratic licence than pertains in the secret and highly classified worlds of threat-based security (Omand, 2010: 73). However, this does not imply the wholesale transfer of sovereign authority to technocrats or the private sector, even if this is where much technical cyber security expertise lies. As Dunn Cavelty (2017: 90) notes, 'the state is not giving away power and authority to private actors' but is 'increasingly enforcing its authority in cyberspace, which has been almost exclusively dominated and shaped by private actors'. This may involve centralization of policy formulation, such as in the UK Cabinet Office, which is the principal government body for cyber security coordination (Martin, 2023). This occurs not by increasing executive power in response to emergency conditions, 'but by using a rationalising discourse of "risk assessment" and "balance" to link policy areas and bring them under an overarching political-economic project' (Neal, 2019: 250). Cyber riskification is therefore not a de-politicizing move, despite its technocratic veneer, but remains under state purview and strategic guidance. In the UK, for instance, the high-level strategic intention since the late 2000s has been 'to make the UK the safest place to live and work online', which has remained consistent despite shifts in weight granted to cyber security and cyber risk (T. Stevens, 2021).

Cyber security and cyber risk are not fixed categories. Slippage occurs in their definitional and practical boundaries, and objects move between them, influenced by the discursive and material investments of influential actors and their strategic priorities (C. Stevens, 2021). The general pattern is for organizations and political institutions to adopt CRM in addition to more conventional threat-centric 'security controls', with the latter part of an expansive organizational risk management programme. On occasions, the opposite happens. The European Union's recent adoption of a more threat-based posture runs parallel to existing cyber risk and resilience policies and practices, occasioned by growing awareness of state-based cyber threats to EU countries and internal EU institutional changes that have facilitated securitization (Christou, 2019; Pawlak and Delerue, 2022; Backman, 2023). In any situation, the objects of risk and security are identified and constructed as risks and threats, thereby enabling them to be addressed in public and corporate policy, with significant variation in emphases across national and corporate contexts (Romaniuk and Manjikian, 2021).

The dual purpose of riskification – and the risk management practices that follow – is to reduce uncertainty about the world through calculation, and to enable the construction of 'risks' that can be governed. As with securitization, riskification can be an elite-driven process, even if its priorities are often with the mundane. The difference is that it is easier to normalize endemic cyber risk than exceptional cyber insecurity. Formulations of the 'new protective state', the democratic polity that seeks to uphold rather than suspend civil liberties in times of uncertainty, are predicated on the notion that government policy exists 'to reduce the risk, so that people can go about their daily lives freely and with confidence' (Mottram, 2008: 49). This appeal to state benevolence masks the ways in which the protective functions of risk might be subverted or misdirected. Risk politics is as prone to elite capture and the possibilities of socially counterproductive or malign ends as security politics (Aradau and Van Munster, 2007; Dillon, 2007). All governance, including cyber risk, has the potential to marginalize, suppress, exclude and eradicate subjects and communities via calculative rationalities that make people and things legible as governable objects.

Conclusion

This chapter has described how cyber security has become augmented, and in some ways subsumed, by a complementary process of riskification. This is a distinctive way of reading and acting in the world that involves the construction of dangers and hazards as risks, thereby enabling the governance of those risks through the ways and means of CRM. CRM differs from cyber security in several ways, the most important being that risk is oriented to the object of protection, its character and vulnerabilities, whereas security concentrates on protecting that entity from known threats. This chapter has also pointed towards the political possibilities of CRM and its potential as a field of governmentality. The following chapter addresses this proposition in more detail and outlines the conceptual approach to governmentality that will be adopted throughout the book. In the transition to the next chapter, we shift from thinking about cyber risk in the round to the specifics of its analysis.

THREE

Analysing Cyber Risk

The academic literature on risk is vast and extends across many different fields of theory and practice. Risk analysis occupies economists, geographers, environmentalists, ethicists, safety scientists and public policy experts, particularly those engaged with natural and technoscientific hazards and with crisis management (Burgess, 2016). For scholars of security, risk has been explored for two decades as an analytical and explanatory framework for understanding the cognition and behaviour of security actors and the policies and practices they set in motion (Lobo-Guerrero, 2010: Petersen, 2012). Prominent in this work is the identification of risk as an emergent form of governmentality, with analyses exposing the operations of power in novel sociotechnical assemblages such as border security, surveillance and counterterrorism (Bell, 2006; Aradau and Van Munster, 2007; Muller, 2010; Jackson, 2015). A secondary strand of literature argues that risk has replaced security as the principal mode of state intervention in the global management of threats relating to interstate war, intrastate conflict and terrorism (Williams, 2008). Both themes are addressed in this book, unified by an overarching analytical framework, as set out in this chapter, centred on a reading of governmentality as transversal fields of power that produce political effects through the practices of cyber risk management (CRM).

This chapter is organized in two sections. The first addresses the inherent uncertainty of cyber risk, its (un)knowability and the structural uncertainties of the operational environment.

It outlines the relationship between uncertainty and risk, arguing that uncertainty is an ontological feature of cyber risk, rather than, as classical risk analysis would have it, what is left over once risk is calculated and domesticated. This discussion then moves into the primary commitment of the book: the applicability of governmentality to the theorization and interpretation of CRM. It argues that governmentality allows for the identification of knowledges and behaviours operant within the overall cyber risk assemblage and which exert political effects. These are not necessarily driven by sovereign 'top-down' decisions or policy but emerge from multiple actors responding to uncertainty in numerous ways, including mobilizing uncertainty for economic gain. The governmentality of CRM materializes through diverse techniques and technologies that act upon and through individuals and entities, often cutting across established analytical boundaries and categories. The second section discusses this 'transversal' aspect of CRM with specific reference to the theoretical resources of international political sociology (IPS) and its insistence on multiplicity and relationality. It then introduces the four 'families' of CRM addressed in the main book chapters: enterprise CRM, cyber insurance, cyber resilience and geopolitical cyber risk. These categories provide methodological purchase on a complex field of risk management by describing how cyber risk thinking is materialized and exploring the governmental effects thereby produced. Their principal features and internal logics are discussed along four interacting dimensions of analysis – practices, spaces, temporalities, subjectivities – and the relevance of each to the analysis is explained.

Uncertainty and governmentality

Cyber risk professionals are 'managers of unease': at every turn, they make decisions in conditions of imperfect knowledge and uncertainty, yet their expertise is essential in shaping the very conditions for the knowability of risk and security

(Bigo, 2002, 2008). The American writer David Weinberger (2012) advanced the idea that the internet was 'too big to know', arguing not just that the sum of digitized knowledge was too large for any one person or organization to perceive and understand in its entirety but also that knowledge in the digital era, as a decentralized, distributed and networked process of knowledge creation, was messy, provisional and contested. This perspective has affinity with postmodernist scepticism as to the scientific method and the knowability of 'truth' (Feyerabend, 1975; Lyotard, 1984) but is more properly a declaration of the contingent and sometimes fragile nature of knowledge constructed in and through digital environments (Borgmann, 1999). This knowledge – or more accurately, knowledges – is often held together by communities of interest more than by stabilized formulations of what constitutes truth in conventionally linear and 'scientific' terms. Cyber risk is the product of sociotechnical processes of knowledge construction: digital technologies play crucial roles not only in what we want to know and why but also in what we can know and how (Ashenden, 2021; Cristiano et al, 2023). Interconnectivity and interdependence are not the abstract notions often encountered in excitable accounts of digital transformation and virtualization but fully materialized entanglements of humans and nonhumans (Aradau, 2010: 509). A growing critical literature indicates that cyber security is 'made' from sociotechnical assemblages that afford, facilitate, constrain and shape its landscapes of possibility (see Liebetrau and Christensen, 2021). In the construction of cyber risks, individual actors will have differential visibility of phenomena and operate under conditions of variable knowability, each shaped by sociotechnical factors, constraints on agency and mobility and by a matrix of competing intentionalities, capabilities, epistemic biases and assumptions. The construction of cyber risks is contingent upon sociotechnical, political and cognitive contexts, each of which expresses significant ontological and epistemic uncertainty.

The conceptual relationship between risk and uncertainty is intellectually fraught. Chicago economist Frank Knight, whose work in the 1920s is foundational to modern risk analysis, bracketed out uncertainty as an unmeasurable residual, distinct from 'measurable risk' that is amenable to probability modelling and calculation (Knight, 2009). For Knight, however, uncertainty is also an inescapable fact of social life and economic decision making. Unless one embraces uncertainty, financial gains (or losses), for instance, will not follow. Knight distinguished between the 'perfect rationality' of ideal risk decision making and the 'intelligence' necessary for responding to uncertainty: '[I]n attempting to act "intelligently", we are attempting to secure adaptation … we do strive to reduce uncertainty, even though we should not want it eliminated from our lives' (Knight, 2009: 124). Nor can we. As Knight suggests, uncertainty inheres in all economic, social and political processes, including those of CRM. It cannot be excluded or written off as a surplus or excess of risk thinking; it is integral to it (Aven, 2025), with important implications for how we understand cyber risk and its management. Uncertainty is also, as well established in the cyber security literature, a condition of the digital environments in which CRM interventions take place (Fouad, 2025b). This arises from the architecture of cyberspace and from imperfect information about actor intentions and the state of digital networks; it is both a structural and epistemic challenge to cyber security decision making (Gomez and Villar, 2018; Brantly, 2021).

At the same time, uncertainty is also a productive resource. In contrast to the top-down, often sovereign, mobilization of risk as a form of control, uncertainty spurs actors' bottom-up (protean) improvisations and adaptations in response to uncertainty (Amoore, 2013: 10–12; Katzenstein and Seybert, 2018: 4). Coupled with a dynamic operational environment and societal change, there is still something experimental about cyber risk, as actors adapt to novel circumstances and try innovative approaches and practices, many of which exploit uncertainty as a means to generate economic returns. Uncertainty is often regarded as an incalculable residual, left over when objective risk has been quantified and tamed, but we need to treat uncertainty as intrinsic to, rather than separate from, the activities of cyber security and cyber risk. It is a condition of their existence – a feature, not a bug, of the environment, as information security professionals might say.

How might we analyse the operations of CRM in such an uncertain and complex environment, and identify its political effects? We can identify two principal inspirations for the theorization of risk in security studies and international relations (Petersen, 2012). The first is the foundational sociological work of the 1990s and 2000s, which identified risk as a dominant motif of thought and action in late capitalist modernity (Giddens, 1990, 1991; Beck, 1992; Luhmann, 2005). These analyses differ in significant respects, but their central proposition is that risk thinking emerges from societal transformations that have destabilized faith in science and reason as the vanguards of liberal progress. As conventional patterns of social, economic and political organization fail to deliver on their promises, leading to the erosion of traditional authority, and as populations look out on a world plagued by shared dangers and threats, risk emerges as a way of rationalizing and managing uncertainty, doubt and anxiety. For Ulrich Beck, technological development brings new hazards, such that the 'risk society' is no longer confronted just with age-old problems of how to survive in a world of perilous natural processes and phenomena but also with the effects of its own technologically induced actions. Moreover, this is a globally integrative process, in which peoples everywhere are bound together in a 'world risk society', a new cosmopolitanism formed by common exposure to financial and environmental hazards, and to the transnational effects of terrorism and other forms of political violence (Beck, 1998, 2009). Crucially, as the future cannot be foreseen except as a domain of fuzzy uncertainty, we are forced to accept a certain ignorance about our technological predicament: we shift from responding to known threats (security) to a risk lens that anticipates possible futures, including those with potentially systemic effects (Goldin and Mariathasan, 2014; Pettersen, 2016; Renn et al, 2019). From this vantage point, for instance, liberal interventionism in the form of peacebuilding and conflict stabilization is a way of managing risk in international society, lowering risk to powerful actors by reducing turmoil (threats) on their peripheries (Clapton, 2014). Even war, viewed through this lens, is an exercise in risk management (Shaw, 2005; Heng, 2006; Rasmussen, 2006; Coker, 2013).

The second draws on critical risk literature and to the techniques and technologies of governmentality, as proposed in Foucault's extensive writing and lectures on the topic (Burchell et al, 1991). Foucault's interest was less with overarching ideologies of risk than with how risk was constructed and operationalized as a calculative rationality of political governance. The central feature of governmentality is 'the conduct of conduct', which means 'any attempt to shape with some degree of deliberation aspects of our behaviour according to particular sets of norms and for a variety of ends' (Dean, 2010: 18). This approach attends to the techniques and technologies deployed by capable and intentional actors to shape the conduct of conduct within social fields of power and knowledge. For Foucault, 'nothing can function as a mechanism of power if it is not deployed according to procedures, instruments, means and objectives which can be validated in more or less coherent systems of knowledge' (Foucault, 1997: 61). Risk is one such system of power-knowledge, constructed and maintained by risk professionals and political elites, within which risk practices are enabled, legitimized and materialized as modes of governance. Risk is thereby identified specifically as a form of governmentality (O'Malley, 2016).

Coercive powers exist in the governmental assemblage but are complemented by non-coercive measures such as education, surveillance, norms and incentives that lead to the willing participation of individuals in social, economic and political activities aligned with the overarching objectives of a relevant authority. The internalization of norms and expectations invites people to play active roles in their own self-governance, often without direct state intervention, which explains why governmentality is often identified with the individualism implied in neoliberalism (Lemke, 2011). Although Foucault's original analyses centred on the state as the primary driver of governmentality, there is significant scope for incorporating other actors and agents as sources of governmental power. There is no a priori reason studies of governmentality should begin and end with the state, as the 'conduct of conduct' can be initiated by and take hold through a wide range of non-state actors and even nonhuman things (Walters, 2012: 145). Inter- or supra-state organizations, overly political or otherwise, also act as 'global governors' (Avant et al, 2010), instantiating transversal fields of power and authority that can be defined as 'global governmentality' (Stetter, 2021). Governmentality beyond the state is of central importance to the current book and its embrace of the distributed and sociotechnical nature of cyber security and cyber risk. Ontologically, governmentality implies plurality, multiplicity and relationality, as it implicates complex webs of actors and intentions that interact in ways that constrain and enable agency and actions. The governmental impulse may derive from higher-level political or strategic imperatives, but the modes of governance are distributed and decentralized across actors and institutions and mediated through sociotechnical assemblages like the internet and society. Power, in the context of the governmentality of cyber risk, is not merely a question of state authority, discipline and coercion but of the pervasive, productive and decentralized ways in which individuals and populations are governed and self-govern.

The governmentality approach theorizes risk as a political technology, in which risk is constructed and operationalized through the techniques and technologies of governmentality to organize and direct social and political life (Lobo-Guerrero, 2010). In Foucauldian terms, risk is a '*dispositif*', a complex assemblage of interconnected practices, institutions and knowledges that form the conditions of possibility for security governance and sociotechnical ordering (Whiting, 2020: 50–4). Diverging from the more overt modalities of state coercion, risk also functions in more subtle yet no less significant ways. Luis Lobo-Guerrero's (2011) analysis of insurance – a key risk technology – for instance, demonstrates how insurance categorizes and capitalizes human life within a political economy of risk that creates new forms of subjectivity and exclusion. Robert Deuchars' (2004) analysis of international finance links the micro practices of accountancy and auditing with the macro-level promotion of neoliberal self-interest masquerading as objective technocracy. Risk governance operates across national and international contexts as systems of power-knowledge that are legitimized within complex assemblages of state and non-state actors while constantly seeking out new targets, vectors and opportunities. This is politics in an everyday register, as well as in the formal arenas of parties and governments: a politics of uncertain knowledge operating in uncertain environments. To address the governmentality of CRM, the following section outlines the book's methodological approach and the dimensions of analysis that structure the empirical chapters.

Methodological framework

The breadth of cyber security and cyber risk is astonishing, as befits a growing global industry and field of national and international policy. This book makes no claims to present an all-encompassing survey or generalizable theory of cyber risk and CRM but instead seeks to make an initial contribution to the problematization of cyber risk. The analysis presented here emerges from a way of reading security, risk and governmentality that aligns with elements of the ongoing project of international political sociology (IPS). IPS emerged principally from European scholars dissatisfied with traditional state-based approaches to security and insecurity, the intellectual Anglophone origins of which – notably strategic studies and international relations – were deemed hegemonic and reproductive of 'symbolic violence' through their persistent reliance on 'self-evident' yet unexamined binaries such as state/society, agency/structure, internal/external, security/insecurity and North/South (CASE Collective, 2006; Basaran et al, 2016). Arguing that these dichotomous framings, and the theory and analyses that proceed from them, 'fracture' the 'reality' of political practices, including risk and security, IPS prioritizes the tracing of practices as they are instantiated in the world, not as theory assumes them to be (Huysmans and Nogueira, 2016, 2021).

IPS stresses the multiplicity and 'transversality' of risk and security practices, which cut across established binaries and disciplines and construct new boundaries and territories, new fields of practice that have constitutive and governmental effects that cannot be captured by conventional readings of national and international security. In this way, it resists the totalizing tendencies of political thought to divide the world into 'levels of analysis' – individual, state, international – that mask the realities of how security and other forms of politics are instantiated in practice (Bigo and Walker, 2007; Hoffmann, 2021). This requires sensitivity to how politics makes 'many worlds' not one, set against the effacement of the local and specific by appeals to 'the global' or 'the planetary', for example (Walker, 1988; Bartelson, 2009). IPS also helps to expose the assumptions and biases of the influential sociological theories of risk outlined previously, which are often rooted in Eurocentric conceptions of modernity and methodological individualism (Deuchars, 2004: 86–99).

This orientation has significant applicability to studying the governmentality of CRM (see Böhm, 2025). CRM is a heterogeneous field of ideas and practices, many of which rely on transversal moves across established institutional or organizational lines. This is not to deny the importance of state actors, firms or international organizations as conventionally understood, but it is to insist on the multiplicity of their relations with other actors within the framework of an ongoing performance of those relations. Any categories – cyber risk included – that emerge from their activities are provisional and contingent and required sustained effort to make them legible as 'risk objects' and capable of effecting change (Van Loon, 2002: 53–8; Power, 2007: 4). Uncertainty is 'priced in' to this way of acting in the world and of interpreting it too. The worlds of cyber risk are not the way they are because of the preordained identities of its operators or any linear notions of causality and inevitability; they are the way they are because relevant actors make them so, under conditions of epistemic and structural uncertainty. Employing an IPS interpretation of governmentality allows for a deeper appreciation of cyber risk as a complex suite of strategies, technologies and techniques that influences the behaviour of people and populations within a broad field of power and cyber risk knowledge.

As a stepping-off point for analysis, we need to disaggregate cyber risk into elements that share the recurring patterns and rationalities that establish its governmental effects. The empirical chapters of the book explore four principal modalities in the CRM landscape: enterprise CRM, cyber risk insurance, cyber resilience and geopolitical cyber risk. There is much internal heterogeneity in each of these categories and important overlaps between them, so they are here treated as 'families' or suites of correlative practices that constitute temporary spatiotemporal stabilizations of fast-mutating fields of cross-cutting thought and action. The purpose is to describe the empirical heterogeneity of CRM in multiple contexts and involving a range of actors with different aims, objectives, techniques and obligations, not to ascribe to each category permanence or inviolability. The intention is not to reassemble these families of practice into a monolithic entity called 'cyber risk', nor to reduce CRM to ideology (neoliberalism) or to market forces (capitalism); words are not equivalent to practices, which in turn do not predetermine desired effects or efficacy. Governmentality is ontologically plural, 'a heterogeneous space, constituted through multiple determinations, and not reducible to a given form of knowledge-power' (Collier, 2009: 99). The aim is to address cyber risk in its diversity and trace the transversal lines that connect agents and structures, and which cut across boundaries and scales. Without this move, we cannot understand how CRM is rationalized and constructed, how its forms interact and how those overlaps and interactions reinforce one another or generate new modalities of risk.

Of interest to the current project is how cyber risk modalities are viewed not just as landscapes of risk but as horizons of possibility and the realization of value: how does CRM operate within the political economy of risk and uncertainty that constitutes the overarching rubric of cyber risk? There are outstanding empirical questions, too, about how CRM channels discriminatory impulses or generates discriminatory effects (Dwyer et al, 2022). If CRM creates new spatiotemporal arrangements, then these are also uneven configurations, where risk and uncertainty can be manipulated and redistributed for political and economic purposes (Shires, 2020; Hassib and Shires, 2021). A transversal approach to CRM shows how risk and uncertainty come into being and presents a novel opportunity for understanding governmentality operating in and through digital spaces.

Each family of CRM is analysed with reference to four dimensions: practices, spaces, temporalities and subjectivities. These dimensions emerge from the concern of governmentality with the rationalities that identify a problem as a problem, its subsequent strategic conversion into an object of risk, the techniques and technologies (practices) developed to address it and the spatiotemporal implications and subjectifications that follow. These dimensions may alternatively be framed as 'logics', 'intersubjective modalities or approaches that draw upon shared material and symbolic resources, carry pronounced conceptualizations and assumptions, and can be detected through discourses and practices' (Backman and Stevens, 2024: 2445–6). They are not isolated in theory or in practice but commingle and coproduce one another, so their deployment here is heuristic and provisional.

Practices. If we accept that risk is 'an essentially contested concept' (see Gallie, 1955), its precise definition is less important than examining 'the role of managerial and administrative practices organized for the explicit purpose of representing and handling risk' (Power, 2007: 3–4). The notion of practice in relation to security and politics has attracted substantial academic theorization and disagreement (Kratochwil, 2011; Lechner and Frost, 2018). Practices are understood here as 'socially meaningful patterns of action, which, in being performed more or less competently, simultaneously embody, act out, and possibly reify background knowledge and discourse in and on the material world' (Adler and Pouliot, 2011: 4). Practices are 'patterned actions that are embedded in particular organized contexts and, as such, are articulated into specific types of action and are socially developed through learning and training' (Adler and Pouliot, 2011: 5). The making of risk, like the making of security, is 'a matter of routine and reproduction', often at the level of 'everyday' practice that incorporates discursive and material resources in its attempts to influence and shape the world (Bueger, 2016: 132). It is also 'relational', in that the management of risk operates between actors acting in a pragmatic fashion that is contingent, reflexive and context specific (McCourt, 2016). The analysis in this book attempts to delineate how the techniques and technologies of cyber risk practice act transversally across boundaries and domains, with reference to specific actors embedded in sociotechnical assemblages.

Spaces. Spaces are a form of ordering, in which actors – through discourse and practice – create spatial arrangements that express their aims and desires but also enable, even demand, governance in order to persist (Massey, 2005). Spaces like the internet are heterogeneous, fluid and contingent; so too the governance arrangements that emerge, which exhibit aspects of both control and protean power. As Collier (2009: 99) observes, governmentality is a heterogeneous, topological space of knowledges and powers, characterized by a 'dynamic process through which existing elements, such as techniques, schemas of analysis, and material forms, are taken up and redeployed, and through which new combinations of elements are shaped'. Topology connotes both the physical topologies of communications networks and the logical topologies of network space and the protocols that organize and facilitate it (Kaufman, 2017: 85). Where risk works through these novel assemblages, it engenders new spatial forms that work alongside the conventional spatialities of the international political system and usher in new forms of sociotechnical governmentality (Walters, 2012: 53). The task of our analysis is to demonstrate how CRM practices reorder the world in spatial terms. Like security, risk is produced in and through sociomaterial assemblages that have both centripetal and centrifugal tendencies: security and risk enclose and reaffirm their spaces of operation but are also open and dispersive, seeking new opportunities for securitization and riskification (Campbell, 2019). The world making of cyber risk is never complete, as new spaces of cyber risk are produced and brought into the ambit of managerial practice.

Temporalities. Conceptually and practically, spatiality cannot be separated from time and temporality. Shires (2019a: 25) identifies the emergence of a 'productive' cyber risk *dispositif* that 'colonises ever further geographically and into the future, endlessly expanding because there are always residual risks', and which constitutes 'an infinite logic where once one risk has been commodified and mitigated, another is identified'. This neatly summarizes the spatiotemporal fluidity of the political economy of cyber risk and its self-perpetuating character in response to uncertainty about the future. The future of cyber risk is 'open', as it is undetermined, but anxieties about the future – more properly, futures – act back reflexively on decision making in the present and vice versa. Time is not an inert metaphysical background to human behaviour and decision making but something that is mobilized and shaped by the practices of CRM (Cristiano et al, 2023). Time is actively 'timed' in any practice that seeks to influence how change occurs (Hom, 2020: 33), and there are distinct temporalities at work in all CRM. The locus of decision making for each is in the present, but the decisions taken – and the practices each entails in the present – articulate different visions of the future and the necessity for specific future actions in each case, albeit in analytically different ways.

Subjectivities. Cyber risk as a political technology produces new forms of personhood and socioeconomic organization. In contrast to the binaries of security – friend/enemy, self/other – risk is less about the direct production of identities than ensuring people's ongoing membership of risk-pooled cohorts, so as 'to actively circulate in the general combinatorial and transactional economy of contingency formed by risk' (Dillon, 2015: 116). This is risk as *bio*political technology, a mode of governmentality embracing ontological contingency and deploying calculative rationalities beyond the state to achieve political and economic objectives (Dillon, 2008; Dillon and Lobo-Guerrero, 2008). Risk insurance has a long history of smoothing out the contingencies of social life, protecting populations and generating economic returns for its providers (Lobo-Guerrero, 2011). It also creates categories of uninsured and uninsurable, new subjectivities created through their exclusion from insurance regimes (Duffield, 2007; Barrinha and Da Mota, 2017). The ubiquity of digital technologies means that all end-users are subjects of CRM: what we can and cannot do in the networks is shaped by technical modulations of cyber risk. Training and education programmes turn us into agents of risk management, which, however apparently benign, shift the burden of everyday cyber security on to individuals (Jarvis, 2019; Nyman, 2021). This is central to cyber resilience, with its accent on responsibilizing individuals, initiatives contested and resisted elsewhere in security (Vaughan-Williams and Stevens, 2016). Important, too, are the many people employed in CRM, charged with implementing complex risk techniques and technologies. Risk professionals are not just subjects of the 'repressive power' of organizational risk management that reproduces 'unequal, contentious and unstable' power relations within organizations (Hörnqvist, 2010: 162) but agents of productive powers that exploit the opportunities of risk and uncertainty for organizational gain.

Conclusion

CRM consists in sociotechnical configurations that extend across sectoral and national borders and which instantiate modes of governmentality through the rationalities, strategies, techniques and technologies of CRM. This chapter has suggested that the primary modalities of cyber risk can be analysed across four dimensions of practice, space, temporality and subjectivity. Taken together, these can begin to describe and explain the structuring effects of CRM practices and their political implications. The intention is not to reify families of CRM as static entities but to highlight their political effects through a provisional analysis of the relationality and transversality of what Foucault (1977: 26–7) once called a 'microphysics of power'. In this framing of governmentality, multiple actors and rationalities interact to exert pervasive influence on dispersed populations of organizational and individual subjects. Cyber risk is constructed, addressed and sometimes sustained, the latter on account of its productive political and economic possibilities within the political economy of cyber risk and cyber security. The arguments raised will be returned to variously in the following chapters, each of which corresponds to one of the four families of CRM practice introduced in this chapter. We turn now to the first of these: enterprise CRM.

FOUR

Enterprise Cyber Risk Management

The term 'cyber risk management' (CRM) is most familiar to practitioners as a set of principles, behaviours and outcomes associated with sustaining the life of organizations. CRM is the process of identifying, assessing and addressing risks to information systems and data and linking these actions to the overall risk profile and welfare of a specific organization. Commonly referred to as 'enterprise' CRM, it is equally applicable to private- and public-sector organizations, all of which face hazards and dangers that are constructed as risks in order to manage them through the strategies, techniques and technologies of CRM. All organizations create, collect, store, process or use data and have some degree of digital dependency that creates vulnerability to external and internal cyber security threats. 'External' threats from criminals, hackers and state-sponsored operations are common and transnational, as globalization increases organizations' exposure to risks originating in the global environment. Even an organization that thinks of itself as 'local' will depend on cross-border supply chains and data traffic routed internationally through multiple jurisdictions and regulatory regimes. 'Internal' threats arise from purposive attempts to subvert an organization from within – disgruntled employees, leakers, whistle-blowers – and from the inevitable failures and accidents of complex IT use, including mistakes by users and by those responsible for configuring and maintaining business systems. Security controls can mitigate these challenges, but the overall risk profile must be managed through a holistic organization-wide approach to CRM. CRM differs between public and private sectors, with the former tending towards more compliance-driven approaches accountable to public regulators. Firms, which have to factor in shareholders and customers, are often more agile and emphasize revenue, reputation and competitive advantage. Organizations will also differ within sectors, as they choose to adopt specific elements of CRM according to their needs and priorities. Of the four families of CRM presented in this book, though, enterprise CRM is its 'purest' form, as it draws most on established risk management practices. It is closest in spirit to the long heritage of technical risk management in other fields and overlaps extensively with cyber security and its primary protective function. There are many ways of framing and presenting CRM, and this chapter presents a synthetic description of an idealized approach to enterprise CRM.

Building on the framework outlined in Chapter Three, this chapter looks at the practices, spaces, temporalities and subjectivities of enterprise CRM. The first section outlines the rationales and practices of CRM with reference to risk assessment and risk management. It introduces the indispensable discrimination between inherent and residual cyber risk and the practices implemented to identify, assess and manage them. The second section describes how CRM creates new spatial configurations within organizations, such as through new functions and technologies, and situates the organization within a wider sociotechnical milieu created by material infrastructures, corporate practices and international standards. The following section develops a temporal account of CRM, in which multiple future-oriented techniques are employed, albeit primarily to enable decision making in the present. The discussion concludes by arguing that CRM brings into being subjectivities – practitioner, end-user – through the shaping effects of organizational priorities and practices on their conditions of existence and opportunities for exercising agency.

Assessing and managing cyber risk

An organization's cyber risk profile is a function of the generic insecurity of digital technologies and data, an organization's attractiveness as a target and its approach to security and risk. UK policy has prioritized raising awareness of cyber risk and promoting CRM tools and guidance for public- and private-sector use (National Cyber Security Centre [NCSC], 2023b). The underlying principle is that cyber risk should not be considered an 'IT risk' but a business risk: poor CRM affects the operational, legal, financial and reputational aspects of an organization (Evans, 2019: 16–17). CRM is not just a 'tick-box exercise' in compliance, or the operational responsibility of IT and security teams, but a core strategic function of senior management. NCSC has developed, for example, a 'Cyber Security Toolkit for Boards', intended 'to help board members govern cyber risk more effectively' (NCSC, 2023a). 'Effective' here means enabling sound decision making that drives security and resilience and creates organizational value. Uptake and implementation of CRM frameworks are variable; they remain dependent on executive understanding of cyber risk, and the resources furnished to drive organizational change (Sloan and Warner, 2020). As risk professionals are keen to note, the design and implementation of CRM are notoriously patchy (Thomas, 2023).

An IT-centric approach to cyber risk maps an organization's technical exposure to cyber risk, arising from vulnerabilities in software, hardware, management systems and digital behaviours. On its own, this would be a reactive stance to events, flaws and errors, leading to post hoc interventions and resource allocation. Proactive CRM looks deeper into organizational dependencies and augments sensitivity to vulnerabilities with assessment of the value of an organization's digital assets. Dependence on an asset gives rise to considerations of its social, political and cultural value: 'Nothing is worth protecting or even saving if it does not have value' (Burgess, 2007: 482). A central focus of risk management is on 'outcomes that have an impact upon what humans value', whether positive or negative (Renn, 2008: 2). Commercially, value is usually understood as economic value that can be diminished or enhanced depending on the efficacy of security and risk controls. Specific forms of risk – operational, reputational, legal-regulatory and third-party/vendor risk – all translate into financial risk. The first step (risk assessment) establishes the business value of digital assets and the costs of business interruption, data exfiltration (for example intellectual property, customer data, corporate strategy) and legal-regulatory implications (Ruan, 2019). This is followed by design and implementation of measures to protect and enhance business value (risk mitigation).

Cyber risk assessment identifies an inventory of systems, technologies, processes and data and the value of these assets to the organization. These can be ranked by relative criticality in an asset hierarchy, with the 'crown jewels' at the top: critical assets like industrial control systems, bulk customer data or transaction platforms, without which an organization cannot function. Other digital assets are ranked by the estimated cost of their failure or disruption, including those for whom the impact may be minimal or zero. Assessment is sometimes differentiated into 'component-driven' and 'system-driven' approaches (NCSC, 2023c). Component-driven risk assessment focuses on individual technical elements of organizational systems; system-driven risk assessment on how these components interact and contribute to overall purpose or function. The issue of value is submerged at the component level but is clearer in system-driven risk assessment, where boards are more involved. Cyber risk assessment generates knowledge of digital inventory and its dependencies; defines baseline metrics of vulnerability through quantitative and qualitative techniques; establishes tolerances to specific risks; and enables decision making about what measures will minimize exposure to cyber risk.

Assessment establishes a level of 'inherent' cyber risk that pertains prior to the implementation of risk mitigation measures. 'Residual' cyber risk is the risk that remains after those measures have been implemented. Security controls comprise the principal response to inherent cyber risk and include logical (firewalls, encryption, intrusion detection, authentication), physical (on-site security, CCTV, employee identification) and administrative (security policies, procedures, training, incident response) practices. Once security controls are implemented, residual cyber risk can be assessed and decisions made about how to manage it, such as transferring cyber risk through insurance mechanisms, or adopting cyber resilience measures that can mitigate risk in a more future-oriented posture. CRM is commonly said to cover both preventive (security) and resilience measures (Falco and Rosenbach, 2022: 8–9).

Risk managers use diverse techniques and technologies to quantify and model cyber risk, communicate risk issues to decision makers and enable the management of inherent and residual cyber risk. Quantitative methodologies enumerate asset values and vulnerabilities. Heat maps, dashboards, risk matrices and other representational technologies 'visibilize' (Amoore and Raley, 2017) risk assessments to help decision makers understand and act on cyber risk. Hardware and software combine in automated algorithmic assemblages to monitor and respond to shifts in security states, increasingly driven by artificial intelligence (AI) and machine learning. CRM operates in a web of sociotechnical sense and decision making (Aanonsen and Knudsen, 2025) that generates risk knowledge through multiple modes of knowledge production and in collaboration with human and nonhuman agents of data collection, organization and interpretation (Lakshmi et al, 2021).

Standards frameworks are also technologies of risk management, with organizations adopting one or more formal CRM framework depending on organization type, context and requirements. These include the influential US National Institute for Standards and Technology (NIST) Cybersecurity Framework, the International Organization for Standards'

ISO 27000 series on information security management and bottom-up frameworks driven by open-source community expertise, like FAIR (Factor Analysis of Information Risk) and OISRU (Open Information Security Risk Universe). Firms also develop and deploy proprietary CRM frameworks tailored to clients' needs. This ecosystem of guidelines, frameworks, standards and best practice is complex and interacts with the broader legal and regulatory landscape, including privacy and data protection, digital resilience and critical infrastructure protection. A major part of the consultancy industry helps clients navigate this field and select the appropriate tools to help them assess and manage their cyber risk exposure.

Deterritorialization and reterritorialization

The primary sites of CRM practice are organizations that have determined that CRM is a necessary and appropriate course of action. Setting aside for one moment that the identification and provision of CRM functionality is more likely to come from outside an organization than from inside it, how does CRM create and order space? The first move is the self-construction of an entity as situated within a spatial (and cognitive) category of risk that is geographically dispersed and universal (Barnard-Wills and Ashenden, 2012). Most risk professionals would agree that cyber risk is a global phenomenon that affects organizations large and small; it is the responsibility of organizations to recognize this 'objective' fact and to respond accordingly via the classic risk mechanisms of accept, avoid, transfer or mitigate. This initial decision locates the organization within a global 'landscape' of cyber risk (see for example World Economic Forum, 2025b), incorporating actors and infrastructures in the horizontal terrestrial plane, below the oceans (subsea cables) and in orbit (European Union Agency for Cybersecurity, 2025). Organizations are therefore nodes in a spatially volumetric (Elden, 2013) matrix of cyber risk. The spatial perspective is dual: looking outwards to global threats and inwards to the conditions of possibility established by an organization's character and configuration.

The second move is the internal reordering of organizations. In well-resourced contexts, this may require the creation of new corporate functions, or the reordering of old ones, to enable the assessment and management of cyber risk. There may be physical relocation of personnel to new operational locales, but this is principally a topological change in hierarchy and relationships, with new roles and lines of reporting, prompted by executive direction and supported by resource (re)allocation. The classic example is the Chief Information Security Officer (CISO), who has responsibility for identifying, assessing and mitigating cyber risks to an organization and interacts directly with organizational leadership (Da Silva, 2022). The CISO function pre-dates the emergence of cyber risk as a widely accepted business risk, but its mandate now includes CRM as standard, with ancillary risk functions fulfilled by specialist teams within an organization. One such is the Security Operations Centre (SOC), which occupies a unique physical and affective space within organizations (Mathew, 2024) and is integral to the implementation and maintenance of the security controls that comprise the primary mitigation of inherent cyber risk.

SOCs and other assemblages of cyber security technologies and expertise contribute to a third move in the reordering of organizational space. Traditional information security models were predicated on the perimeter defence of organizational networks and generated inside–outside mentalities that sit uneasily with the reality of most cyber security and cyber risk (Betz and Stevens, 2013). Perimeter defence is all but impossible, on account of the transit of information across organizational boundaries as a condition of digital interconnectivity, and the increased use of cloud services, remote working and other offsite corporate activities. Each makes network boundaries hard to secure, as the inside–outside demarcation is stretched and blurred, creating new geographies of data exchange and use. In the early 2000s, 'de-perimeterization' emerged as a new model of network security that accepted the porosity of organizational boundaries, leading to more refined approaches to data management and to device and user authentication like 'zero trust' architectures (Spencer and Pizio, 2024). De-perimeterization took a risk-based approach to organizational cyber security that worked as much from the inside out as the traditional threat-based model worked from the outside in. De-perimeterization has not occluded the classic inclusion–exclusion binary of security, but it has led to the creation of liminal spaces in which users, devices and data outside the physical integuments of organizations are enrolled within the (topo)logical structures of those organizations (Vuorinen and Uusitupa, 2022). In turn, the organization is transformed spatially, as its operations extend beyond its physical location and into spatial zones, like the home, not formally under its control or purview. None of these dynamics is unique to cyber risk, but their trajectories are interlaced with the emergence of CRM and the de- and reterritorialization of organizational topologies and the dispersal of labour.

The centrifugal spatialization of cyber risk is illustrated by the early appreciation that risk-based de-perimeterization provided economic opportunities for organizations – by improving their risk posture and capacity to reach into transnational markets – and for third parties providing the necessary services and tools to do so (Spencer and Pizio, 2024). Many organizations will buy in security and CRM capabilities, as they are unlikely to have developed them in-house. This complicates organizational geographies, as risk advisors and consultants move in and out of organizations, security functions are outsourced or offshored and technologies are provided and maintained by third-party contractors. Risk considerations ripple across organizational boundaries and up and down lengthy supply chains, especially, in the latter case, for manufacturing firms that depend upon transnational suppliers, or which are pivotal nodes in transversal assemblages of economic exchange and value creation (Huang et al, 2021; Jones, 2025).

Significant portions of the CRM economy are concerned with the implementation of standards and regulatory compliance. CRM standards have been developed notably in the United States (NIST) and the International Organization for Standards (ISO), and then exported globally, albeit with an admixture of bottom-up standards initiatives. NIST CRM standards, for instance, have a specific geographic derivation (US), as do standards like FAIR and OISRU developed in the Global North. The material documents disseminated by these organizations 'convey standards across space with minimal distortion' and also produce forms of spatiality individuated by the internalization of procedures and protocols (Dittmer, 2017: 84–5). CRM standards bind corporate practices into spatialized territories within which 'proper' implementation takes place and outside which it does not. These are not strictly geographical – adoption and deployment are uneven within any political-territorial unit – but emerge from Global North centres of influence and authority and are dispersed through transnational networks of capital and corporate interest. There are limits to this dispersal, as countries like China and Russia require compliance with domestic standards that mirror rather than import NIST or ISO, although individual companies within those countries may adopt international standards to access global markets that recognize them as best practice (Saalman et al, 2024). Firms within and outside these countries must navigate dual streams of standards-based regulation that map in one case to local priorities and in the other to international frameworks derived principally from the US and Europe.

Organizational CRM situates organizations within a global cyber landscape, reconfigures their operational topologies and is sustained by a global economy of CRM that produces new territories of authority and influence. These de- and reterritorializations are enabled by the sociomaterial underpinnings of global information infrastructures, upon which organizations rely for their essential activities, but also through networks of capital and the less visible workings of international standards. Cyber risk has transformed the conventional binaries of threat-based security into more elastic and contextual spatial configurations that refuse the traditional aporia of security – its illusion of universality (Burke, 2002) – and refer instead to the mobility and fluidity of risk and insecurity.

Decision making in the extended present

The primary cognitive orientation of CRM is to the future: like all risk practices, it exists to determine the likelihood and harm of future adverse events. As a concept, risk is 'essentially a temporal one, grounded in its relation to an unknown future … [it] expresses not something that has happened or is happening, but something that might happen' (Reith, 2004: 386). Risk, as it emerged during early modernity, was a way of quantifying the probability of future events, based on aggregating data over the long term and identifying patterns of occurrences that rendered future uncertainties calculable and controllable (Hacking, 1990). This allowed for the development of financial instruments to commoditize risk and thereby profit from uncertainty (Ewald, 1991; Bernstein, 1996), a persistent feature of contemporary global political economy. Developing in lockstep with scientific theories of mechanistic reality and causal determinism, risk implied that humans could have agency in the future, previously a temporal realm beyond human control or understanding. As Reith (2004) articulates, however, this future 'openness' to human agency was undermined by the technologies of late modernity that compressed space and time and brought instantaneity and acceleration to causal relationships. Under these conditions, 'the future' effectively disappeared as a discrete arena of human intervention and was subsumed within an 'extended present', in which 'concerns about the future – in fact, the construction of the future itself – dictate the present, not the other way around' (Stevens, 2016: 93–4). In this temporal regime, risk transforms from a technology of probability to one of possibility (Amoore, 2013), in which risk serves decision making in the present rather than in the future.

We can see these theoretical arguments play out in CRM. Cyber risk analysts note that the operational environment is determined by the 'speed of data' (Evans, 2019: 5), the relentless exchange of which reduces the value of reactive postures (security) and requires proactive (risk) management instead. In this formulation, the future (and the danger) is already here, on account of the velocity and volume of information enabled by global information technologies and the practical eradication of conventional understandings of cause and effect. There can be no waiting around to deal with future problems if the problems are already within the organization in the form of vulnerabilities being exploited by adversarial malware, for instance. While this orientation underplays the continuing importance of cyber security – as we have seen, this is one way of mitigating inherent cyber risk – it indicates an important shift to the temporal present as the principal site of concern and of decision making. Cyber insurance and cyber resilience are more future oriented in that they accept that future harmful events are highly likely, if not inevitable, but the prime function of CRM is decision making in the present.

A vital facet of the temporality of CRM is the anticipation and imagination of possible futures, which construct cyber risks as objects to be governed in the present through the practices of risk management. An assessment of existing vulnerabilities in an organization's IT configuration or business practice is irrelevant until it is harmonized with an understanding of what the harms (operational, financial, reputational) will be to an organization if those vulnerabilities are exploited. These are future oriented, expressed as what could or will happen should remedial actions not be taken in the present.

This may be a trite observation: all decisions are taken in the present, including decisions to make decisions later (or not at all) (Osborn and Simpson, 2018). The important point here is that these speculative futures act back on the present reflexively. Whereas data about the past can certainly be used to inform risk decision making and the capitalization of future uncertainty, CRM also generates data about the future to drive decision making in the present. These data are not classically empirical but speculative or possibilistic, produced by a range of quantitative and qualitative methodologies that attempt to model likelihood and harm in the absence of data about future specific events. The technical artefacts that visibilize and communicate cyber risks to decision makers are sociotechnical translations of possibilities into legible forms (Hall et al, 2015). They stabilize uncertainty, however temporarily, to allow for the application of calculation as a precursor to decision making about how to manage specific and overall cyber risk profiles. These may, of course, be quite accurate – what is the direct financial hit if our data centre is unavailable for 36 hours? – and can be augmented by patterned historical and experiential data, but these risks are always uncertain and unknowable in strict terms.

CRM is always caught in an extended present that exhibits two key characteristics. The first is the need to manage and counter known security issues through the implementation of technical, physical and administrative security controls in real time. The second is a register of anticipation and apprehension that imagines future and unknown dangers in order to construct them as risks to be accepted, mitigated, transferred or terminated, often through other practices like insurance or resilience (Backman and Stevens, 2024). In both cases, decisions are made in the present: frequently automated in the case of security controls and the mitigation of inherent risk, more often strategic when dealing with residual cyber risk. The full 'extension' of the present, however, is more fundamental still, as the CRM project is never complete.

Decisions are just punctums on risk management journeys that last for as long as an enterprise exists and which are buffeted by changing environmental conditions and managerial priorities. Interviews with CISOs indicate their perennial frustration with this ontological condition of uncertainty and the impossibility of attaining levels of risk reduction that are even 'good enough' to satisfy CRM objectives, which also change over time (Da Silva, 2022: 156–7).

Identity and agency

Understood as business risk, CRM's principal locus of decision making and responsibility is the executive board and its ancillary functions. Risk assessment is a necessary component of risk posture but is insufficient without operationalization of its findings by organizational leadership (Falco and Rosenbach, 2022). Strong executive commitment leads to efficient management structures, appropriate delegation of responsibilities, distribution of resources and clear lines of communication and reporting. Corporate leaders organize and oversee dedicated CRM programmes; cyber security policies and controls; assessment, auditing and monitoring; education, training and awareness in service of 'cyber security culture'; incident management; and public relations. They aim to reduce cyber risk to levels acceptable to the organization – and its stakeholders, if applicable – sufficient to sustain or enhance enterprise value and discharge its fiduciary and regulatory responsibilities. Enterprise leaders are the most visible agents of CRM and primary subjects of its adoption, operationalization and management. Leadership is responsible for developing CRM strategies, within which cyber risk is articulated in specific ways congruent with the aspirations of the organization. In this fashion, cyber risks are 'strategy-specific' (Hörnqvist, 2010: 20), and leaders direct the internal life of an organization through cyber risk frames that shape the conditions of action for those who work within it.

At its simplest, CRM has required the creation of roles dedicated to its professional practice that did not exist ten or 20 years ago: cyber risk and compliance manager, cyber risk analyst, cyber risk consultant, IT-risk and compliance manager, IT-security and risk manager, to name a few, plus a bewildering array of job titles in technical cyber security (UK Cyber Security Council, nd). These subjectivities are enrolled within the CRM assemblage and collectively constitute the labour pool for delivery of an overall cyber risk strategy. As the strategy is paramount, cyber risk is communicated within an organization and over time as a form of 'intra-organizational power' (Hörnqvist, 2010: 161) that guides professionals towards organizational objectives and targets. This is cyber risk operating as a form of governance in its own right (Baker and Simon, 2002), as it sets the discursive and material boundaries of subjects' agency, while ensuring its own reproduction. Individual practitioners may challenge the nature of their roles, express their scepticism about CRM methodologies (Stevens, 2023b; Simpson, 2024) or even admit confusion about their professional identity (Da Silva and Jensen, 2022), but CRM persists as a strategic endeavour that smooths out local irregularities in service of higher-level objectives and helps to 'negate alternative epistemologies' (Deuchars, 2004: 171).

At a more detailed level, the sociotechnical nature of CRM creates new forms of human and nonhuman subjectivity and agency. Professionals work closely with technical devices to generate cyber risk knowledge and through which decision making is distributed in an 'assemblage of subjects and objects' (Amoore, 2013: 162). High-level decisions – such as overarching strategy – are principally taken by humans, but the conditions of possibility for decision making are often created by lower-level processes that are markedly computational and automated. This is strikingly the case in the security control contexts that aim to reduce inherent cyber risk. Cyber security has used machine learning, and now 'AI', for years in security operations, including threat detection and anomaly identification, vulnerability management, automated incident response and profiling of system and user behaviours (Stanham, 2023; Aitel and Geer, 2025). High data volumes and operational efficiency demand the use of algorithmic decision-making agents to analyse and order data at speed, generating intelligence that can feed into other organizational decision-making loops. As a result, 'analysts and operators are enmeshed in hybrid human–machine assemblages that generate knowledge of particular utility in specific decision-making circumstances' (Stevens, 2020: 167). The processes of knowledge construction in these sociotechnical configurations suggests that human and nonhuman subjectivities are co-constituted through relational entanglements that in neither case are purely human or algorithmic (Liebetrau and Christensen, 2021).

An important purpose of CRM's security controls is to manage what end-users can and cannot do within the topological spaces of an organization. It also modulates how external users interact with that organization. Cyber security controls manage users' access to digital systems, data and resources; authenticate user identities; restrict what internal and external resources users are authorized to view or modify; and control what devices and data users can introduce to an enterprise network. Like classical information security, these controls exist to protect the CIA triad – of data confidentiality, integrity and availability – but they are also designed to reduce overall risk to the organization. Some actions may be proscribed – visiting bad web domains while using company networks; using unauthorized data storage media or platforms; use of Virtual Private Networks – but alternative routes are provided where the users' desired activities align with organizational objectives, legal responsibilities and norms of appropriate behaviour. A company has no obligation to permit access to personal email or streaming entertainment in the workplace, but it will find security solutions that enable employees to 'bring your own device' to work or nominate some applications as approved or otherwise for defined forms of organizational activity.

These are practical solutions to improve security and reduce risk, but they also shape the conditions of possibility for user behaviours. This need not be interpreted negatively: few employees deliberately wish to cause harm to their employer; nor would they object to security measures that make an organizational network safer and more productive to navigate. Most users are not even aware of the invisible work of this facet of CRM, even if it prevents their desired objectives. It also barely registers in the substantial academic and policy literature on the 'human factor' in cyber security, which is interested more in raising awareness of how end-users can improve cyber security (Khando et al, 2021) than in the effects of security controls on users. It arises obliquely in discussions of poorly designed security affecting employee productivity through 'security fatigue' (Stanton et al, 2016) or incentivizing unauthorized and potentially insecure 'workarounds' (Blythe et al, 2013), but the effects of 'good' security on the construction of organizational subjectivities are rarely discussed. Successful technical measures as part of overall CRM attempt to influence user behaviours and reduce their agency within accepted limits: not to prevent them from acting at all but to shape the paths of action open to them. End-users are enlisted nodes within a risk–security assemblage that configures – often dynamically – who can do what, when and with whom, a matrix of sociotechnical relationality that shifts according to organizational risk priorities as well as assessments of external and internal threat.

Conclusion

Enterprise CRM is a complex and evolving set of practices that attempt to reduce the risk to an organization of a range of security threats and negative outcomes. It reconfigures organizations in space and time, brings about new agents and subjects of risk and, with future wellbeing in mind, enables decision making in a turbulent present. It operates in and through shifting sociotechnical constellations of hardware, software, data and people, which it shapes in subtle yet recognizable ways. The spaces produced by enterprise CRM are international and global – through mobile standards operating transversally across national borders, policy and the development of a transnational support sector – and also hyperlocal in implementation. Depending on perspective, CRM either smooths out irregularities on a global scale, or looks increasingly fractal as the aperture narrows to focus on its specificities and internal diversity. Empirically and conceptually, it must be both, even as we can identify commonalities of thought and practice that cohere CRM across its sites of implementation. The principal function of enterprise CRM is to enable decision making about how to address inherent and residual cyber risk. Inherent cyber risk is reduced by security controls and other forms of mitigation, yet residual cyber risk will always remain even after these measures have been implemented. The following chapter addresses a crucial technique for transferring residual cyber risk – cyber insurance – and its functions within the assemblage of CRM.

FIVE

Cyber Insurance

Cyber insurance is a growing but immature field of cyber risk transfer. The first standalone cyber insurance policies emerged in 1997 in the United States, but a minimal client base, inconsistent approaches to risk valuation and coverage and exogeneous events like Y2K (the so-called Millennium Bug) and the dot.com collapse of the early 2000s meant that it was not until the 2010s that awareness and uptake of cyber insurance became a mainstream business concern (Camillo, 2017; Wolff, 2022: 43ff). The value of the global cyber insurance market was estimated at $15.3 billion in 2024 with the potential to double by 2030 (Munich Re, 2025). The principal rationale of cyber risk insurance is protecting the insured from financial losses and liabilities arising from cyber incidents. The need for cyber insurance arises because firms cannot – or will not – absorb financial or other impacts caused by future negative events. The fundamental mechanism of insurance is that organizations transfer cyber risk to a third party (the insurer); they pay now to recoup losses later. Increasingly, insurers will perform a second move, with most cyber risk insurers transferring all of part of their risk to specialist reinsurers to manage their own financial exposure in case of large or catastrophic cyber events (Skeoch and Ioannadis, 2024; Woods and Wolff, 2025: 8). Cyber insurance will not stop incidents from happening, but it may help deal with and recover from incidents if they occur during the period covered by the insurance policy.

This chapter outlines the main elements of cyber risk insurance through examination of its practices, spatialities, temporalities and subjectivities. The first section sets out the main categories of first- and third-party cyber insurance that manage residual cyber risk, as well as the ways in which insurers help customers address inherent cyber risk. It introduces the notion of insurance as a form of governance, in which insurers intervene directly in organizations and shape the political economy of cyber risk. The following section examines the spatial features of cyber insurance, such as its concentration in the markets and traditional insurance centres of the Global North. It highlights the impelling logic of cyber insurance that searches out profit through the transversal exploitation of market opportunities, thereby imparting both aggregational and dispersive characteristics to the spatialization of cyber insurance. The third section discusses the temporal aspects of cyber insurance, with specific reference to the commoditization of the future, notions of operational and financial normality and the 'strategization' of time through the imagination of future events. The last section explores notions of subjectivity through the lenses of 'insured' and 'uninsured' entities and returns to the theme of 'uninsurability' that runs throughout the chapter.

Underwriting as risk governance

Organizations that have implemented security controls to mitigate inherent cyber risk as an element of cyber risk management (CRM) may seek to transfer residual cyber risk to a third party through cyber insurance. First-party coverage extends to losses incurred directly by the insured party, such as through business interruption and revenue loss; third-party liability covers costs generated by litigation, regulatory penalties or other actions taken by a party against the insured subsequent to a data breach or other harmful event. Third-party dynamics are further complicated by the interconnectedness of the operational environment and the systemic cyber risk that arises through enterprise interdependencies, which means that cyber risks are rarely confined to a single organization (Khalili et al, 2019).

In the absence of longitudinal sets of reliable incident data that enable actuarial calculations of probability and loss, it remains difficult to underwrite cyber insurance products at levels that protect customers and insurers (Romanosky et al, 2019; Skeoch and Ioannidis, 2024). Insurers face the persistent problem of establishing viable levels of cover for a wide range of first- and third-party cyber risk classes, which include 'loss or damage to digital assets, data recovery, business interruption, notification costs, liability in respect of data breaches, multimedia liability, employee dishonesty, cyber extortion and regulatory defence costs' (Camillo, 2017: 55). All cyber insurers need to triangulate coverage, solvency and data collection issues to ensure profit in normal years and service their contractual commitments should large claims occur (Woods and Wolff, 2025: 2–3). The heated discussion around whether harmful state-sponsored cyber operations constitute *forces majeures* – unforeseeable or catastrophic events that permit suspension of insurers' stipulated obligations – illustrates nervousness about offering coverage of low-frequency, high-impact, perhaps even systemic incidents whose costs might run to billions of dollars (Wolff, 2024). This has prompted proposals for government reinsurance backstops akin to those for terrorism, although the potential systemic or catastrophic effects of cyber incidents question the ability of reinsurers to price and carry cyber risk (Cremer et al, 2024a; Scroxton, 2025). Mismatches between customer expectations and policy wording, especially in a technical field like cyber security, also favour the use of exclusion clauses, coverage reduction or non-payment of claims (Cremer et al, 2024b).

Of the most common categories of coverage, the top four are what might be called 'clean-up' costs incurred directly by the insured, such as breach notifications, regulatory or other costs, services to affected individuals inside and outside an organization and, importantly, public relations (PR) services (Romanosky et al, 2019: 7). If an alternative practitioner framing of risk is 'risk = hazard + outrage' (Sandman, 1989), management of anger, shock and indignation following an adverse cyber event is a central function of today's enterprises. Companies that have failed to recognize the importance of proactive PR to limit 'outrage' have suffered at the hands of customers, shareholders, media, regulators and the market, leading to grave damage to their earnings and reputation (Boakye et al, 2024). Cyber insurers will underwrite the costs of PR and crisis management, as with other loss categories, and also provide other 'value-added' services, such as incident response, forensic investigatory capacity, security and risk management guidance and legal support, or connect them with third parties offering the same (MacColl et al, 2021).

As insurers are willing only to cover residual risk, they are keen to establish whether and to what extent customers manage inherent cyber risk. Many of the value-added services offered by insurers address inherent risk by offering tailored consultancy, advice and guidance for managing it. They may also undertake auditing of new customers' security controls and continual monitoring of clients' security postures throughout policy lifecycles, although this is not always the case (Baker and Shortland, 2023). While understanding that the primary purpose of cyber insurance is risk transfer, not the improvement of cyber security per se (MacColl et al, 2021), it is in insurers' interests to motivate better cyber security as it reduces residual risk and therefore insurers' risk exposure. This is cyber insurance practice as 'loss prevention', in which insurers intend to prevent losses occurring in the first place by intervening directly in the life of the insured (Baker and Shortland, 2023), analogous to a fire insurer providing free smoke alarms to its clients. 'Loss control' is also evident, intended to minimize the severity and impact of losses when they do happen, through the provision of incident management, post-event mitigation and training support (Lillen, 2025). Underwriting practices unfold in the body of the insurer – client profiling, actuarial data analytics, risk assessment – but, in the form of data collection increasingly enabled by automation and artificial intelligence, are interwoven with the organizational CRM and security functions of the insured too.

This reach-back into organizational practices and decision making has led to cyber insurers being described as de facto regulators (Camillo, 2017) that shape and modify the internal life of organizations through underwriting practices and the imposition of standards and policy. The technical details of cyber insurance markets and products are less important for the current analysis than the framing of cyber insurance as a form of governance that shapes the cyber risk environment and the political economy of cyber risk through direct intervention into organizational life (Baker and Shortland, 2023). This is not to suggest that cyber insurance is a 'silver bullet' for cyber security or cyber risk governance (Adriko and Nurse, 2025). In practice, insureds fail to align their security postures with standards like ISO 27001, and insurers ignore non-technical aspects of those standards – like corporate reporting and accountability mechanisms, employee awareness and training and incident response planning – that are essential components of holistic and effective CRM. If cyber insurance can be read as a form of governance, it is uneven, sometimes ineffective and can only help mitigate cyber risk as part of a wider suite of security and risk practices (Mott et al, 2023).

Aggregation and dispersal

Cyber insurance, through the mechanism of risk transferral to third parties, has an inherently spatial dimension. Cyber risk is distributed beyond the confines of organizational boundaries within a transnational insurance market that shifts residual risk from organizations to insurers. This process of 'making uncertainty fungible' (Lobo-Guerrero, 2011) has created a multi-billion-dollar global cyber insurance market directly employing tens of thousands of people and affecting the working practices of hundreds of millions more.

The label 'global' implies a smooth distribution of cyber risk and standalone cyber insurance, but the cyber insurance market is unevenly distributed (as well as internally variegated). Like other aspects of the services market, the international political economy of cyber insurance is heavily skewed towards North America and Europe. According to Swiss Re (2024), a major insurance and reinsurance provider, North America accounts for 70 per cent of global cyber insurance premiums, followed by Europe (19 per cent) and the countries of the Asia-Pacific (8 per cent). The remainder is accounted for by the Global South, and by China, which has a small but growing indigenous cyber insurance market (Gallagher Re, 2024). The current situation reflects the historical dominance of the traditional centres of insurance (London, Zurich, New York), with an admixture of smaller hubs (Tokyo, Singapore, Hong Kong), the relative maturity of cyber security markets and regulatory frameworks in those regions and entry issues for Global South actors with differential access to the financial resources necessary to purchase cyber insurance coverage.

The strong implication is that cyber insurance coverage reproduces established corporate insurance geographies whose centres of capital, authority and ideation are firmly situated in the Global North. This suggests the construction of an insurantial imaginary (Ewald, 1991: 198) that demarcates the 'inside' of the neoliberal and risk-managed Global North from the more uncertain – so, 'risky' – 'outside' of the Global South. This is partly true but cannot be reduced to simple cartography for two reasons. First, cyber insurers are cautious in extending cover to Global South countries due to their less robust cyber security infrastructure, lack of historical data and, in cases like Latin America, an appreciation of significant contemporary rises in cyber incidents (Hardy and Blanc, 2023). Second, and for the same reasons, these regions present market opportunities for cyber insurers, if they can help prospective clients manage their inherent cyber risk profiles, develop tailored insurance policies, improve data collection and advance the development of tighter cyber security and cyber risk regulation. The spatialization of cyber risk insurance is contingent on the gradual intervention of market and regulatory forces, and the insertion of sociotechnical techniques and technologies into local settings, rather than the imposition of cyber risk insurance frameworks on emerging economies (Herr, 2021). International organizations like the Organisation for Economic Co-operation and Development (2017) recommend adoption of cyber insurance as a tenet of national cyber security and resilience strategies but cannot impose this by virtue of their limited mandate. Similarly, industry bodies like the Global Federation of Insurance Associations (2024) view cyber insurance as an important means of addressing cyber risk, although its influence is limited to policy advocacy and lobbying.

This creates a globally complex and fragmented space of cyber risk insurance, with a vibrant core in the Global North and tendrils extending rhizomatically into the periphery. Its networked, rather than hierarchical, spatiality depends on the agency of individual insurers and insureds, and their propensities and capacities, and is not determined primarily by legal jurisdiction or political geography, although they overlie and reinforce one another in numerous ways. This networked heterogeneity also characterizes the internal structures of developed cyber insurance markets, such as found in the core countries of the Global North. These markets are far from uniform, with disparities found between sectors and organizational types, and between individual countries' policies and practices. Small and medium-sized enterprises (SMEs), for example, while constituting 90 per cent of all businesses, are less likely to purchase cyber insurance than large companies (Adriko and Nurse, 2024). The reasons include reduced SME risk awareness and resource constraints, as well as insurers' lack of historical data, which affects their ability to price coverage effectively. Low SME insurance uptake has been identified as a factor in increased systemic cyber risk, as the coupling of low CRM capacity with SMEs' ubiquity in the economic landscape could result in widespread effects should a cyber incident spread through relatively unsecured networks (Adriko and Nurse, 2024). Countries vary widely in their approaches to cyber insurance too, with some advocating strongly for organizations to seek suitable coverage, whereas others adopt a more cautious approach to cyber insurance, concentrating more on essential cyber security as a precursor to developing a cyber insurance market (Lemnitzer, 2021). There is significant variation in the supply and demand for cyber insurance that maps partially to national contexts but is also transversal across political borders and economic boundaries.

Insurance is one strategy through which risk 'allows the transformation of uncertainty into matter amenable to trade and exchange' (Lobo-Guerrero, 2011: 123). Through processes of risk transfer, residual cyber risk is commoditized and transferred to a third party that bears the costs of a future negative event in return for the payment of insurance premiums. Cyber risk is, in effect, moved from one locality (the insured) to another (the insurer). Insurers concentrate capital in specific locations – the balance sheet of a major firm headquartered in the Global North, for instance – but they also aggregate cyber risk and therefore their own exposure to future adverse events that can only be modelled and not predicted. Underwriting cyber insurance is practical when risks are restricted to an organization, less so if they spread or are 'correlated' across insureds or become systemic through the networked transmission of nonlinear effects (Eling and Schnell, 2016: 479). Here, cyber insurance becomes difficult or impossible to price as it exposes the insurer to unbearable losses, particularly if multiple insureds claim concurrently – as might happen after a systemic cyber event (Khalili et al, 2019). Insurers may seek to exclude classes of cyber risk from their policies entirely, like the 'war exclusions' discussed previously. The spatial concentration of residual cyber risk may be a factor in reducing coverage of that same risk. Insurers can advise their clients to disaggregate their corporate IT structures, such as reducing nodes of potential failure by not relying on single technology providers and dispersing data centres and cloud services geographically (Pain, 2023: 25). They can also diversify their own portfolios to balance risk distribution and manage their own exposure to cyber risk (Cremer et al, 2024a: 9).

This discussion illustrates that cyber insurance performs hidden functions in the spatialization of cyber risk, as the managerial and financial landscapes of insurance intersect with and produce new configurations of sociotechnical infrastructure. Cyber insurance operates through the calculation and commodification of risk, sometimes leveraging extant economic geographies, at other times constructing new ones as insurers search out opportunities and markets. Through a combination of networked and hierarchical operations, market and state interventions, insurers aggregate capital and risk in a heterogenous assemblage of actors and technologies entangled topologically in a dynamic environment that is both global and local. Like other forms of insurance, cyber insurance transforms old spatial orders and constructs novel arrangements with distinct spatial features.

Commodifying contingency

Through a combination of actuarial and other practices, insurance constructs categories of cyber risk so as to render them calculable and commodifiable within the political economy of cyber risk. As an instance of governmentality, insurance is concerned with 'taming the future' (Aradau and Van Munster, 2008a, b), a 'moral technology', in which the calculation of a risk is 'to master time, to discipline the future', such that 'even in misfortune, one retains responsibility for one's affairs by possessing the means to repair its effects' (Ewald, 1991: 207). The connection between insurance and responsibility is implied frequently in cyber insurance commentary and advice – you 'need' it because it provides a 'financial safety net', is 'proof of due diligence on behalf of governing bodies' and gives boards 'peace of mind for themselves and their stakeholders' (Schueler, 2024). Enterprise leaders have agency in whether or not to purchase cyber insurance, but it is the responsible – and responsibilizing – course of action to do so if circumstances and resources allow it. The contingencies against which cyber insurance is constructed are future 'events' – data breaches, ransomware attacks, system failures – and cyber insurance is explicitly situated as a means to ensure business continuity, by which must be meant the continued pursual of economic gain and sustenance of capital, and the preservation of social order. In this respect, the precautionary principle guiding insurance functions to ensure 'the indefinite continuity of an unmodified present, the noninterruption of social and economic processes and normality' (Aradau and Van Munster, 2008a: 201). Cyber insurance functions to return entities to baselines of operational and financial normalcy under conditions of uncertainty.

The development of cyber insurance has been a story of how events previously thought uninsurable have become insurable objects. There is still significant disagreement about how to insure against particular eventualities – ransomware incidents being a conspicuous example (MacColl et al, 2023) – but the insurance industry has been adept at developing new products to cover new classes of events. This 'entrepreneurial power' emerges from the innovative agency of corporate actors, who, '[s]eeking to secure against the uncertain … venture into areas for which knowledge is patchy, at best, and chart the spaces that were until then considered uninsurable or catastrophic' (Lobo-Guerrero, 2011: 10). Actuarial calculation alone cannot propel this task, so insurers, like other cyber risk professionals, have to rely on other cognitive resources and modes of knowledge production. Imagining future events is built on 'patchy' data (probability) and on intuition, expertise and creative modes of thinking and practice (possibility). Through the construction of future events as events, and therefore as objects of insurance, this element of the cyber risk *dispositif* again illustrates its productive capacities and its propensity to colonize the future – there will always be residual risk; the question is how to categorize and commodify it in the pursuit of profit (Shires, 2019a: 25).

Two classes of future event are relevant to the discussion of cyber risk. The first is the relentless daily onslaught of what we might call 'minor' or low-level cyber security events, caused by the probing of firewalls and endpoints, phishing attempts and anomalous network behaviours. These are the principal emphases of security controls and protective cyber security. The second class of event is more significant, comprising minor events that have escalated into something more consequential, or standalone operations that have serious impact on an enterprise network, like ransomware, major data exfiltration or loss of business function. Cyber security has often called into question the wisdom of focusing on major events at the expense of the mundane (Cristiano et al, 2023: 337), but in cyber insurance major incidents are the principal object of insurability. Cyber insurance can advise on 'minor' events – loss prevention through security controls – but it can only insure against 'major' events that have potentially measurable loss profiles and can be constructed as aesthetic and instrumental objects in their own right (Stevens, 2016: 151–2). Future events are constructed – imagined, modelled, simulated, evaluated – as objects of insurance but, more importantly, are performed through the practices of cyber insurance. The construction of a future populated by cyber events is an active process of ordering the future (Hom, 2020) so as to act upon those events in the present. Lobo-Guerrero (2014: 366) describes this, following Foucault, as a 'strategization of time', in which insurers 'liberate insurability from the temporal strictures of traditional actuarial practices and create an infinite space for market development'. Insurers do not predict risky events with precision but use them as vehicles for the mediation and negotiation of financial objectives.

The 'quasi-infinite longevity' created by insurance arrangements contains a dual bargain between insurers and insureds (Ewald, 1991: 209). Insurers wield product innovation as a means to commodify the future, but they are also expected to pay out when insureds seek compensation for a covered loss or damage. By its nature, insurance projects insureds' expectations into the future, so the unwillingness of cyber insurers to compensate for some classes of event has caused consternation among insureds and in the insurance industry (Forscey et al, 2022). The aforementioned disputes about cyber *forces majeures* arose when insurers refused to meet what insureds believed were their contractual obligations with relation to high-impact events, indicating that it is not just the modelling of future events that is uncertain but the content of cyber insurance policies (Meland et al, 2017; Tatar et al, 2021). The insurance industry is working to reduce interpretive ambiguities – and legal challenges – by adapting its approach to *forces majeures*, including through proposals for state reinsurance mechanisms (Lockton Re, 2025). This ushers in the potential role of the state as the ultimate guarantor of cyber insurance, as we find already in fields like terrorism reinsurance, and which are posited for systemic and catastrophic cyber events whose effects will be felt far beyond the confines of individual organizations. Public–private cyber risk pools through cyber reinsurance schemes equivalent to Pool Re (terrorism) or Flood Re (flooding) do not yet exist, but their possibility indicates that the futures of society and state may become bound up with cyber risk as a biopolitical technology in years to come.

Insurability and uninsurability

The potential convergence of private cyber insurance with state reinsurance indicates that public–private boundaries may be eroding in cyber insurance. The public–private nature of cyber security has been a consistent focus of analysis and policy, informed by widespread understanding that the operational environment does not permit simplistic division into public and private domains (Carr, 2016; Collier, 2018). However, even prior to state-backed reinsurance proposals, the private world of cyber risk insurance has always been entangled with public authority. Insurance policies are legally enforceable contracts and can only exist within coherent legal systems. The absence of specific 'cyber insurance law' may be unhelpful in failing to improve and harmonize cyber insurance markets, but governments have been legally instrumental in shaping those markets (Wolff, 2022) and maintain the legal backdrop against which insurance operates in the first place. This indicates, from a governmentality perspective, the interpenetration of sovereign power and authority (law) with more entrepreneurial, productive modes of power (insurance) (Lobo-Guerrero, 2011: 127). These forms of governmentality differ in how they engender subjectivity: law attends to the individuality of the potential insured, its characteristics and behaviours; insurance measures the individual entity relative to specific classes of events against which they are to be insured (Lobo-Guerrero, 2011: 129).

It matters to the law how many people an SME employs or whether it has paid its taxes. These facts are less relevant to the cyber insurer, who is only concerned with the SME's risk profile in relation to particular future possibilities. Whereas a government may identify that SME as a specific entity in law, insurers will treat it as multiple subjectivities, each constructed in relation to specific insurable events (Lobo-Guerrero, 2011: 130). A cyber insurer may insure an entity against one class of event but not another, dependent on its calculation of the specific risks to that organization. The same entity may be insured in one area but uninsured in another, a bifurcation in identity that is sustained in insurance but not in law. Organizations have agency in this process, although it is delimited by insurers. They can, with the help of insurers or other sources of advice and guidance, including national regulators, manage down their residual risk to insurable levels by the implementation of security controls. This is agency in a positive valence, as actors determine how to achieve their goals through affirmative action. In a more negative register, organizations might, if the residual risk remains too high for an organization's risk appetite, decide to avoid the risks of a subset of IT-related activities (Young et al, 2016), a difficult proposition for any enterprise that depends upon digital infrastructure, goods and services. Some organizations may decide not to obtain coverage at all, adjudging it prohibitively expensive, leading to what one major reinsurer has called a 'huge cyber protection gap' in the SME market (Swiss Re, 2024). In these cases, actors withdraw from classes of activity, or from the insurance market, as responses to the expectations of insurers and so pursue different routes to being uninsured.

Our earlier discussion noted the disparity between cyber insurance coverage in the Global North and Global South or, more accurately, between high-income countries with ICT-dependent economies and societies, and lower-income countries with less well-developed ICT infrastructures and socioeconomic dependencies. It is tempting to present this situation as evidence that insurers view the latter as 'uninsurable' due to institutional capacity and legal-regulatory gaps. This critique applies to other areas of security (for example Barrinha and Da Mota, 2017), but the logic of the cyber insurance market suggests that these regions are not uninsurable but under insured. They represent untapped markets for which insurers will develop products once data availability and structural conditions allow the calculative modalities of insurance to produce cyber risks in these areas as legible, governable and profitable. The subjects of cyber insurance are always potentialities, nodes of future value that become enrolled in the cyber insurance assemblage as circumstances permit. Populations of uninsured actors 'represent a condition of possibility for insurance practice' (Lobo-Guerrero, 2011: 125), and, once a threshold of viability is reached, potential value can be realized as actual value.

The spectre of uninsurability is not so easily dispelled, however. Setting aside actors whose risk profiles are determinedly uninsurable, current discussions of the uninsurability of cyber risk shift emphasis from the potential subjects of insurance to the uninsurability of some classes of cyber risk. As the chief executive of major insurers Zurich predicted of the rising possibilities for large-scale cyber events, 'what will become uninsurable is going to be cyber … this is not just data … this is about civilisation' (Smith, 2022). Referring implicitly to *forces majeures* and cyber events equivalent to war, they argued that it may become impossible to insure against incidents whose harms are experienced systemically at the level of populations rather than discrete organizations. If there is a zone of uninsurability, it is not in low-income countries but in the richer countries of the world, whose societies and economies are deeply interconnected and dependent on ICTs. Were such events to occur, of course, their effects would be felt globally, through disruptions to communications and supply chains, as rehearsed by any number of ICT outages and other forms of transboundary crises (Backman, 2021). Although it is likely that the insurance industry will, again, find new ways of insuring against even the most extreme of cyber events – including through reinsurance – the idea that highly capitalized countries may be uninsurable has significant implications for those societies and for perceptions of national and economic security (Johansmeyer, 2023). It would exclude specific events from insurance coverage, as well as populations and organizations expecting insurance against harms far beyond their localized capacities to prevent.

Conclusion

Cyber insurance is a technique of governmentality that constructs future events in order to commoditize cyber risks, both in established and emerging markets, and to make them governable through the products and practices of the insurance industry. This is not to criticize cyber insurance, which has many social and economic benefits, but it is to problematize its underlying logic and imply that further investigations into the practices and orientations of cyber insurance are desirable and necessary. Cyber insurance is still an experimental field of insurance, as the novelty, dynamism and uncertainty of cyberspace are not easily addressed with the wisdom and expertise of insurance practice in other areas. It is becoming clear that there are questions even about whether all classes of cyber risk can be insured effectively, at least through the private sector, and if additional mechanisms of state reinsurance will be required to provide ongoing protection to economies and societies. As residual cyber risk will always be present, even once CRM and insurance have worked together to reduce inherent and residual cyber risk, another form of risk management takes on noteworthy salience, which differs again in its assumptions and expressions: cyber resilience, the subject of the following chapter.

SIX

Cyber Resilience

Neither cyber risk management (CRM) nor cyber insurance will prevent the materialization of inherent or residual cyber risks as data breaches, ransomware demands or other forms of crime, disruption, sabotage and accident. At some point, all organizations are likely to be affected by a cyber incident, whether it is deliberately targeted at them or not. In the UK, 42 per cent of small businesses and 74 per cent of large businesses reported an impactful cyber incident in 2024 (Department for Science, Innovation and Technology, 2025). The levels of under reporting, awareness and visibility issues, and other 'missing data' problems (Oppenheimer, 2024), imply that the scale and severity of cyber incidents are likely to be significantly higher. Indeed, researchers suggest that only 3 per cent of incidents are captured in publicly available data sets (Sangari et al, 2022). The 'when, not if' perspective that has long informed cyber security and CRM therefore has empirical justification. The challenge for organizations is to acknowledge that the existence of residual risk means that future impacts are possible, even likely, and to prepare accordingly. Since the early 2000s, 'cyber resilience' has emerged as a conceptual and practical framework for addressing this aspect of residual risk and, while numerous definitions exist (Tzavara and Vassiliadis, 2024), it is broadly understood as 'the ability of systems to resist and recover from, or adapt to, a cyber compromise' (Kott and Linkov, 2021: 80). From an organizational perspective, it can be simplified as 'the ability to continuously deliver the intended outcome despite adverse cyber events' (Björck et al, 2015: 312), meaning the capacity to ensure business continuity. Cyber resilience is now the dominant framing of cyber risk in many countries, with new policy and legislation coming onstream to prepare people and organizations to be resilient when inevitable future events occur.

This chapter analyses cyber resilience in four sections. The first describes the practices of cyber resilience, which are informed by the need to anticipate, absorb, recover from and adapt to negative cyber security events. The emphasis here is on preparedness, which anticipates future events and works towards the cyber resilience that will be required when those future events occur. The second section looks at how cyber resilience is represented spatially through the metaphor of 'ecosystem' and the implications of this framing for negotiating the tensions between localized resilience behaviours and global digital systems. The third section investigates the complex temporalities of resilience, in which resilience links past, present and future in ways that differ from other risk-related forms of policy and practice. It pays particular attention to the role of failure as a necessary driver of adaptive resilience. The last section discusses the multiplicities of the 'cyber-resilient subject', its vulnerability and relationality, and the problematic insistence on the possibilities for agency that accompany the 'responsibilization' of actors through resilience agendas. The cyber-resilient subject is always faced with radical uncertainty, caught between the unknowable but hopefully resilient future and the less than resilient present.

Preparedness and uncertainty

Cyber resilience – like the materialization of risks as tangible phenomena – is only manifest when the subject of resilience has to respond to an adverse event. Until that point, it is discernible as a range of practices intended to foster resilience and through the perception of the capacity of an individual, organization or system to 'perform at an acceptable level' when a harmful event occurs (Hubbard, 2023: 210). If cyber security is about avoiding system compromise in response to a range of knowable cyber threats, or minimizing system downtime when an event occurs, cyber resilience is 'what happens after an adverse event and requires preparedness for both known and unknown threats', underscoring the importance of recovery from future events (Kott and Linkov, 2021: 80). This is at the centre of crisis 'preparedness' agendas, which articulate the preparatory steps necessary now in order to be resilient in the future (see for example HM Government, 2025a). Resilience is conceptually and temporally distinct from classical risk. Risk manages down vulnerability and harm in the face of known threats and unknown eventualities; resilience 'embrace[s] uncertainty about future disruptions and responds to threats that cannot be averted in time' (Kaufmann, 2017: 86). Cyber resilience is quintessentially 'when, not if'.

A lack of resilience capacity is illustrated by countless breaches and compromises that have led to interruptions in business continuity. These are cyber security failures, of course, but they are also cyber resilience failures, in that organizations were unable to perform their business-critical functions and responsibilities once a cyber security compromise occurred. In 2023, essential digital services of the British Library, the national library and legal depository of all books published in the UK, effectively went offline for months following a ransomware incident (British Library, 2024). Described by a former senior government official as 'one of the worst cyber incidents in British history' (Ash, 2024), a subsequent government audit found that a reliance on legacy systems was partly to blame, which affected its resilience capacity and increased its post-incident recovery time (National Audit Office, 2025: 41). British Library management was widely praised for their transparency following the ransomware incident (Martin, 2024a), but fundamental questions remained about the organization's prior resilience posture and preparedness. The same report criticized the UK government for poor cyber preparedness across the public sector: not only did its resilience posture have 'significant gaps', but government had 'not met their responsibilities to improve their own and their wider sectors' cyber resilience' (National Audit Office, 2025: 12).

Cyber resilience practices are varied and cluster around four correlated and mutually reinforcing principles: anticipate, absorb, recover and adapt. Anticipation is pre-event and involves a proactive approach to cyber security that understands digital assets and invests in robust security controls and processes. Technical security is not resilience (Kott and Linkov, 2021), but anticipatory or pre-emptive measures are fundamental to resilience, as they also include planning for when an incident occurs: what to do, by whom and when. Resilience capacity is difficult to develop 'on the fly', so forward planning and preparedness are necessary for determining what actions to take 'when, not if' the worst happens. Absorbing the impact of an event is in part a technical issue, in which cyber security specialists seek to contain 'the degree of degradation' (Kott and Linkov, 2021: 81) of a system, but it also speaks to the ability of organizational leaders to understand what is happening and to move confidently and swiftly to the recovery phase on the basis of incident response planning. This details actions during and after an incident, so that appropriate organizational resources can be mobilized rapidly to isolate affected systems, bring back-ups online, restore data, communicate to stakeholders effectively (including customers, shareholders, markets and the media) and achieve business continuity as quickly as possible.

All phases of preparedness involve exercises that enrol participants in simulations of future crises and test their incident response capacities in line with organizational plans and priorities. Different teams will be required to respond in different ways, so exercises are tailored to specific horizontal functions within an organization (security, communications, financial, legal) and to the relative responsibilities of managers and leaders at various levels in the organizational hierarchy. They are also differentiated by incident and threat types, such as the UK National Cyber Security Centre's (NCSC) 'Exercise in a Box' suite of online resources, which covers data breaches, ransomware, insider threats and other sources of cyber risk (NCSC, nd, b). Regular exercising helps 'to establish a baseline of how resilient you are now and exposes areas for improvement' (Brown, 2024). Adaptation is an essential part of resilience preparedness. Like cyber security and CRM in general, once an event has occurred, post-incident adaptation requires leadership that not only understands the issues at stake but is prepared to invest in resilience and adjust organizational processes and, often, mindsets too (Hill and Creese, 2021). Adaptation requires not just articulation of 'lessons learned' but the practical and conceptual revision of extant practices to cope with any future incidents. Effective cyber resilience marshals social, organizational and technical competencies within an overall framework that recognizes the sociotechnical nature of cyber security, cyber risk and the digital ecosystem (Amir and Kant, 2018; Dunn Cavelty et al, 2023) and the processual and adaptive character of cyber resilience that nurtures 'latent organizational capacities' (Dupont, 2019: 7).

The growing awareness of organizations' exposure to large-scale data breaches, ransomware and supply-chain attacks has helped cyber resilience become a dominant framing for cyber security and CRM in the 2020s (Creese and Joshi, 2024). Organizations often struggle to develop cyber resilience within existing internal arrangements, and a sizeable consultancy industry has grown to support them. Insurers also play a role in cyber resilience by auditing cyber security controls as a precondition for contracts and by offering incident response support and other post-event services (Mott et al, 2023). Governments have moved to regulate and legislate for cyber resilience, such as the UK's forthcoming Cyber Security and Resilience Bill that widens the remit of existing regulation to improve organizational cyber security and resilience. At the supranational level, the legal-regulatory mechanisms of the European Union have put cyber resilience at their core

(Carrapico and Farrand, 2024), and other regional groupings like the Association of Southeast Asian Nations recognize the importance of cyber resilience to their collective well-being (Benincasa, 2020). International standards bodies have also developed specific frameworks for promoting and evaluating cyber resilience (see for example Ross et al, 2021). Cyber resilience is a common feature of national cyber assessment frameworks (NCSC, 2024a) and local initiatives geared to cyber and wider societal resilience (National Cyber Resilience Centre Group, nd). It is also fundamental to cyber security capacity building, particularly in the Global South (Tabansky et al, 2024). These governance innovations recognize the entanglement of organizational cyber resilience with national and collective cyber security (Christou, 2016; Tiirmaa-Klaar, 2016).

Ecosystemic entanglements

The motif of entanglement is indicated strongly in cyber resilience and cyber security, which are often framed with reference to the particular spatiotemporal configuration of the 'ecosystem'. In biology, an ecosystem is a community of organisms interacting with the physical environment, sustained by complex flows of nutrients, energy and water. The term has been translated into other fields to describe networks of interconnected agents, including in cyber security, where it is often semiotically linked with cognate metaphors drawn from biology (viruses, worms) and public health (infection, 'cyber hygiene') (Betz and Stevens, 2013). It is a common formulation in public policy, where cyber security is couched in ecosystemic terms that emphasize the need to harness capabilities, labour, innovation and the market to deliver economic growth and national security (Christou, 2016). Frequently, cyber resilience comes to the fore as the principal desirable quality of these ecosystems, as they develop the capacities to absorb exogenous shocks, continue operating effectively during a crisis and 'build back better' afterwards (HM Government, 2021: 65). Resilience draws on its terminological heritage in the biological sciences, with an additional cybernetic resonance that implies post-disturbance reversion to an ecosystemic equilibrium (Grove, 2018: 171–3). Cyber resilience encapsulates both a descriptive component – this ecosystem is resilient (or not) – and a normative and programmatic one: this ecosystem should be resilient, and this is how we achieve that goal.

The ecosystem metaphor, when applied to the global information environment, implicates a wide range of actors (human and nonhuman), organizations, networks and other sociotechnical entities (Christou, 2016: 35). It also produces a vision of the operational environment that is spatialized globally, cutting transversally across national and corporate boundaries on account of the transnational materiality of physical infrastructure and data flows (Steed, 2019: 15–21). Like other 'generative metaphors' in cyber security, the ecosystem analogy allows for the definition of 'a mental model of the problem that makes certain policy solutions appear natural or appropriate' (Slupska, 2021: 467). The holistic, yet heterogeneous, attributes of digital ecosystems sustain the adoption of a cyber resilience frame that is 'expressive of a specific form of space that necessitates resilience' as a mode of governance (Kaufmann, 2017: 85). The internet, for instance, is a dynamic assemblage of interconnected systems and networks that also displays significant complexity, which in turn may give rise to unpredictable and nonlinear effects that cannot be predicted in advance (Crowcroft, 2010). There is a 'constant interplay of disruption and equilibrium', which the adaptive qualities of resilience are well placed to regulate because resilience 'comprehends insecurity and disruption as its permanent condition' (Kaufmann, 2017: 86). Cyber resilience may intend a return to 'normal baselines' of system behaviour (equilibria) but only within an environment of ontological uncertainty. Resilient systems therefore can never display equilibria in a pure sense but are always in a state of change and nonlinear emergence (Chandler, 2018a: 47).

The impulse behind cyber resilience in these spaces is not to prohibit particular activities but to control and regulate connectivity and fluidity, 'to let local change happen, but to manage it in order to make it productive for the whole system' (Kaufmann, 2017: 91). There is a tension between the ambition to make individual nodes (people, organizations) resilient and the higher-level rules, policies and standards of national or global cyber resilience, as they address different spatialities and political contexts. However, as Mareile Kaufmann (2017: 90) argues, this combination of localism and globalism is 'captured by the ecosystemic rationale that combines local actors into one overarching, complex system that follows a set of harmonized rules', such as the technical protocols that govern information flows and the relations between nodes in the ecosystem. Protocols within networked spaces are productive and governmental, as they are 'both an apparatus that facilitates networks and a logic that governs how things are done within that apparatus' (Galloway and Thacker, 2007: 29). Protocols produce and govern network spatialities, within which, drawing on the resilient ecosystems literature, nodes can self-organize in ways that meet the higher-level resilience goals of the ecosystem, whether this be organizational, national or global. This does not imply that rules do not constrain agency – protocols and standards are often highly politicized (DeNardis, 2009) – but that particular forms of relational agency are enabled by rules operating in and through spatial configurations of information technologies and users.

The role of cyber resilience as a productive or catalytic political choice is registered in national policy. A ministerial statement outlining the UK's approach to cyber security and resilience noted the importance of cyber resilience in 'fostering an environment in which investment and innovation can thrive' (HM Government, 2025b). The ecosystem in this case is the UK, to be further protected by 'essential cyber defences' and conventional cyber security, which is situated in an 'increasingly dangerous and unstable world'. As this is national policy, the UK must draw territorial lines through the transnational operational environment, but the emphasis on resilience in cultivating stability and 'a secure foundation upon which new ideas and technologies can be built' is consistent with a reading of resilience as accepting of uncertainty and generative of economic and social productivity. The role of the UK government – and other rule makers – is 'the creation of opportunities through stimulation and control via modulation, distribution and flexibility' (Kaufmann, 2017: 91) and the exploitation of agency within the spatial integuments of territorial authority.

Understood in ecosystemic terms, the spatial difference between security and resilience becomes clear. Security works by constructing binaries of outside/inside, public/private, state/individual, which often produce or reify exclusionary spatial and sociopolitical orders. In contrast, resilience, through the ecosystem metaphor, operates transversally and holistically in complex sociotechnical assemblages that are inherently uncertain and always in the process of becoming. Resilience creates and modulates spaces in which agents are interconnected relationally, a topological organization of space that reflects and reaffirms the networked character of information systems. Carving out national boundaries for cyber resilience is determined by political necessity, not the operational environment, and allows for the implementation of rules that foment resilience as a means to empower and enable agents to thrive. However, the limits of sovereign power are brought into sharp relief by the inability of cyber resilience to influence all the components of the ecosystem, which extend far beyond national borders and whose dynamics act back upon the territory in question. Like other forms of risk management, resilience can never be finished in time or in space.

Failing into the future

Resilience is the intended outcome of preparedness: it comes into force only at the moment of the event, when the claims that X or Y is resilient are evaluated against the conditions that pertain at that time. There is a twin anticipation at work here: first, in the anticipation of the future event and, second, in the anticipation of the responses to that event. The first is determined – bad things will happen – while the second is less certain and ambiguous. Anticipation is always performed in the future tense: what will happen, or what might happen? Cyber resilience is conducted in this register, as it attempts to predict the character, if not the precise timing, of future events so as to prepare individuals for how they should think and act when they happen. In one influential analysis, this requires the construction of a 'sensorium of anticipation', in which security and resilience practices imagine and rehearse the future event 'in order to render the future palpable and foster subjects who can inhabit the future not just through fear and anxiety but also through desire' (Aradau and Van Munster, 2011: 85–6). Preparedness training and exercises are intended to cultivate resilience and impart the self-perception of agency to participants, as well as inculcate an urge to become resilient in response to future events (Aradau and Van Munster, 2011: 95).

As with the 'invisible' infrastructures that are so often the object of resilience practices and policy, shortfalls in preparing to be resilient are usually only revealed when resilience is needed most (Star, 1999: 382). Failures of preparedness, and hence of resilience, 'reveal the bureaucratic and material histories of objects and the assemblages in which they are networked' (Stevens, 2016: 119), in this case the practices and political economy of cyber resilience and cyber security. In the resilience framework, however, failure is an opportunity to learn and adapt, either from the event itself – the data breach, ransomware attack or system outage – or from the relative success or otherwise of an entity's resilience posture once an event occurs. Cyber resilience might be interpreted as 'the domestication of security failure', in which the embrace of uncertainty and contingency are the rationale for resilience, and for which 'security failure becomes part of the story about security learning and improvements in capability' (Heath-Kelly, 2015: 69). In addition to the anticipation of future events, security failures and other interruptions of normality become historical data points that inform the futures of cyber security and resilience. In effect, resilience policies and practices 'utilize and redeploy past examples of security failure to promise a more secure future' (Heath-Kelly, 2015: 70).

The future of resilience is open ended, and with it the possibilities for political reworking and anxiety. Adaptation in the wake of crisis is a major component of what the UK government calls 'active resilience'. In distinction to 'passive resilience', which is the capacity to absorb stress and return to normality, active resilience means 'growing progressively tougher by learning from adversity and becoming better able to avoid and manage future stresses' (Martin and Giddings, 2020: 4). There are limits to this dynamic, though, as no actor can continue to become 'more resilient' indefinitely. Once a future event has happened, there will always be another, and another, from which actors must learn (again) from failure and become ever more resilient. This is a logical absurdity, or at least has sharply diminishing returns, and the generalizing effect, were resilience to become the dominant mode of cyber policy (as seems to be the case), is that resilience may serve to heighten anxiety about ubiquitous vulnerabilities and the possibilities of harm, rather than diminish it (Evans and Reid, 2014: 91–5). That the reality of future events never quite matches their anticipated character compounds this anxiety as it 'punctures all preconceived fantasies about the foreboding event' (Evans and Reid, 2014: 94) and the value of the preparation that precedes it. The future, even if handled adroitly, continues to be 'uncertain and traumatic' (Dunn Cavelty et al, 2015: 7). Uncertainty about the future becomes uncertainty about the value of resilience, leading to feelings of helplessness and subsequent denial of, or disengagement from, the very adaptive measures intended to increase cyber resilience (Joinson et al, 2023).

Cyber resilience is a more temporally expansive concept than either CRM, insurance or security. If security is situated primarily in the present and CRM and cyber insurance operate principally in a futural register, resilience draws upon the pasts of actual and future events to inform preparedness in the present that pays dividends in the future. It is uniquely reflexive in that it emphasizes adaptation and transformation under conditions of ongoing uncertainty in ways that other risk-related activities do not. These differences inform analyses arguing that cyber resilience is not a form of risk management at all: risk management seeks to minimize hazards; resilience expects hazards and stimulates continued functioning when they manifest materially, especially when they are unknown, unexpected or otherwise beyond the scope of traditional risk models (Dupont, 2019: 2). The ubiquity of failures – of systems, preparedness and future operational states – characterizes the terrain of cyber resilience but would not be tolerated in the world of security or classical risk management. The prospect of failure motivates cyber insurance – in the hope that failures will not occur – but actual failure, partial or otherwise, is a requirement for the persistence of cyber resilience as an adaptive learning process. This is not to argue that failure is the desired end state of resilience, only that failures provide 'turning points' in the ongoing project of cyber resilience (Dunn Cavelty et al, 2023: 811). The lurking uneasiness in cyber resilience, however, as in other fields of cyber security and risk management, is that the systemic collapse of information technologies bound together in the digital ecosystem may overwhelm the capacities of any actor to absorb, recover or adapt. Resilience aims to nurture future agency, but it has its limits. This point of ultimate failure reveals the 'uncontrollability of the world' (Rosa, 2020) and renders *cyber* resilience redundant, shading instead into individual resilience and societal resilience writ large. This sense of permanent vulnerability permeates all forms of resilience, and in the temporalities of cyber resilience we see too that we are always 'delicately poised between past memory and future possibility' (Evans and Reid, 2013: 89), each characterized by uncertainty and anxiety. Cyber security exemplifies this condition well, as in its everyday operations the prospect of the terminal catastrophe matters less than the ongoing experience of persistent insecurity and pervasive risk.

Context and embodiment

The principles of cyber resilience – anticipate, absorb, recover, adapt – work together through practice to produce 'resilient subjects' capable of withstanding and learning from cyber security failures. The resilient subject has been the focus of significant analysis and critique, often arguing that pushing agency downwards from the state to lower-level entities like firms, infrastructures and individuals is a way of devolving political responsibility for security and risk management. In this line of thinking, the 'responsibilized' corporate entity or person is, under the guise of enhancing their ability to act under conditions of stress or turmoil, effectively being abandoned to their own vulnerabilities in a world that is unknowable and uncontrollable (Evans and Reid, 2014). In this reading, the resilient subject is also the neoliberal subject, free to exercise individual agency but only within the constraints of the market and the state (Joseph, 2018). At the extreme, it is argued that we should understand resilience through 'the pernicious forms of subjugation it burdens people with, its deceitful emancipatory claims that force people to embrace their servitude as though it were their liberation' (Evans and Reid, 2015: 154). This critique is unlikely to garner much recognition or sympathy among cyber resilience professionals or policy makers but, if we accept that cyber resilience – as with any other field of risk and security – produces subjects through practice and discourse, what can we say about cyber-resilient subjects and the forms of agency afforded them under conditions of uncertainty?

Any discussion of the production of cyber-resilient subjects must recognize that resilience is not an intrinsic capacity of an entity but a process through which resilience is continually made and remade. Within the web of the cyber security ecosystem, nodes (people, organizations, infrastructures) are in constant interaction with other nodes, the networked contingency of which conditions both the emergence of insecurity in complex sociotechnical systems and the possibilities for resilience (Kaufmann, 2013). A small- or medium-sized enterprise might have cause to feel vulnerable to cybercrime perpetrated through digital networks, but those same networks can provide resources to aid preparedness and resilience, as well as the ability to leverage network forms in self-organizing behaviours that mould to local conditions and needs better than any directives from above. One issue with visceral critiques of resilience is the confusion of resilience as a concept with resilience as ideology, particularly as it expresses and co-produces a neoliberal reading of the world (Bourbeau, 2018: 47–8). Subsuming the former within the latter grants states and other actors the authority to determine what is and is not meant by resilience, which masks the diversity and multiplicity of resilience as it is experienced and practised by actors on the ground. Governments certainly do encourage cyber resilience in forms entirely congruent with neoliberal motives (for example HM Government, 2025b), but how cyber resilience is developed and implemented in practice, before and after the adverse event, is far more likely to be influenced by local context – 'communal history, collective memory and social convention' (Bourbeau, 2018: 51) – than by theory or ideology. Cyber-resilient subjects are multiple, relational and contextual, partly constrained by rules and protocols, but also relatively free to adapt how they adapt, drawing on community resources and social capital in their efforts to weather cyber insecurity and to learn for the future (Jarjoui et al, 2024).

The theme of agency as empowerment is a feature of cyber resilience policy. The UK's National Preparedness Commission argues that the 'simple end goal of government guidance should now be to empower firms to achieve critical operational resilience' (Dimitriadis, 2025). The role of government is to advise and guide entities along the path of cyber resilience, empowering them to determine how they use the material, economic and social resources available to them. In many cases, though, the availability of resources and the agency to exploit them do not mean that improvements in cyber resilience are inevitable. 'Psychosocial' factors like trust and community cohesion interact with material factors to produce a diversity of resilience outcomes that disturb any linear relationship between, for example, affluence and cyber resilience (Dunn Cavelty et al, 2023: 805–6). Even in rich countries, it is insufficient to exhort actors to 'be cyber resilient' if they do not understand either what that imperative means or how to achieve it. In lower-income contexts, the resources may simply not be available anyway (Tabansky et al, 2024). Cyber resilience is deeply context specific and, unsurprisingly, unevenly distributed.

The cyber-resilient subject, despite its agentic opportunities, is an uncertain subject. Like risk, cyber resilience is predicated on the uncertainty of the cyberspace environment and what we can know about it – it is both ontological and epistemic (Chapter Three). The uncertainty of cyber resilience is more radical still, as it is unconcerned with the causes of problems, just their effects (Kaufmann, 2017: 74). Whereas risk manages down uncertainty, resilience must embrace the uncertain as a condition of its pursuance. The cyber-resilient subject is always one of 'embodied uncertainty', in which epistemic and ontological uncertainty combine in complex sociotechnical assemblages to produce cyber-resilient subjects that exist in a persistent state of unknowability and therefore vulnerability (Dunn Cavelty et al, 2023: 809). Cyber-resilient subjects of all kinds do not calculate risk as such but dwell in a liminal space between 'two worlds – the invisible powerful world which is resilient and the apparent world which appears vulnerable to catastrophic irruptions' (Aradau and Van Munster, 2011: 47). The promise of future cyber resilience is always coupled with the uncertain reality of lived experience and produces, through the practices of preparedness and resilience, in principle at least, the 'discovery of new subjective capabilities' (Aradau and Van Munster, 2011: 48) that designate 'being cyber resilient' in adverse conditions.

Conclusion

Cyber resilience is becoming the dominant framing of cyber problems and solutions in public and corporate policy. Leveraging metaphors from ecology and strands of neoliberal political ontology, cyber resilience cultivates agency and self-organization of cyber-resilient subjects such that they can prepare for, absorb, recover from and adapt to the complex and dynamic conditions of contemporary technological life. Failure, vulnerability and uncertainty characterize the conditions of possibility for cyber resilience, without which there would be no need for resilience, nor any grounds to pursue what is an open-ended project extending infinitely into the future. While we can argue that 'resilience is not risk' (it is not), cyber resilience works together with CRM within the ambit of cyber security to enhance security and stability in the IT ecosystem and wider society. Cyber resilience produces governmental effects, seeking not to discipline or disbar but to enable and affirm the possibilities for thriving and flourishing under conditions of radical uncertainty. The role of the state is to validate and empower actors to become cyber-resilient, rather than dictate the precise conditions, valences and boundaries of that resilience. There are open questions about the capacities of particular actors to capitalize upon this invitation, based on differential access to material and immaterial resources. Cyber resilience is unevenly distributed and forever incomplete, which undermines the overall resilience of the greater digital ecosystem.

The previous three chapters have explored specific families of cyber risk and resilience practices and orientations. Much of this discussion has concerned organizations, and in this chapter individuals, with less explicit attention paid to the state and government. Chapter Seven shifts perspective to the state and addresses aspects of cyber risk as they pertain to state attitudes and behaviours. It looks at how cyber risk interacts with geopolitics and how states attempt to manage cyber risk in the international environment.

SEVEN

Geopolitical Cyber Risk

A strictly threat-neutral approach to cyber risk would prioritize how to manage and respond to cyber incidents rather than concentrate on the precise source of actual or potential cyber threats. However, organizations, particularly in strategic sectors or critical infrastructure, are routinely targeted by actors motivated by international politics or who have links to state actors. In these cases, their risk portfolios will include assessments of geopolitical cyber risk, that is, those cyber risks that are identified with the dynamics of global cyberspace and with the ways in which international actors treat 'cyber' as a strategic resource to be acquired and used in strategic competition (Dunn Cavelty, 2025). The combination of the 'new normal' of geopolitical cyber operations (Buchanan, 2020) with the inextricable web of public–private interdependencies in global cyberspace means that organizations are increasingly mindful of geopolitical cyber risk and adjust their risk management practices accordingly. Common responses include revising their insurance provision (risk transfer), shifts in operating policies (risk management and mitigation) and cessation of corporate activities in particular countries or regions (risk avoidance) (World Economic Forum, 2025b: 26). Governments face their own specific challenges, as they identify and address cyber risks that, by definition, extend beyond sectoral or national boundaries. These include cyber risks emerging from geopolitical tensions, inter-state strategic competition, war, conflict and state and state-sanctioned cybercrime. In turn, geopolitical interests hinder the development of collective intergovernmental positions on cyber security and cyber risk (Sukumar et al, 2024). If much of the previous discussion in this book has been about cyber risk as a business risk, this chapter engages with state responses to cyber risks as a form of 'geopolitical risk' that states navigate cautiously but which also offers opportunities for productive action aligned with their national interests (Gould-Davies, 2019; Rodban, 2025).

The first part of the chapter shows how governments reincorporate geopolitical and other cyber threats into their risk management practices, as distinct from more threat-neutral frameworks. It introduces the concept of 'preventive risk management' as a modality through which states attempt to reduce risk by proactively countering and neutralizing known cyber threats, including those associated with ongoing geopolitical competition. The second section examines how preventive cyber risk management (CRM) spatializes cyberspace through the dual move of its universalization as a global domain and its reterritorialization through differentiated topological zones of friendly and adversarial space. It also explores the continued relevance of inside/outside territorial binaries, even as these political constructions contrast markedly with the transversal operational environment. The temporal aspects of geopolitical cyber risk are addressed in the third part of the chapter, which identifies the precautionary principle intrinsic to national risk management and the temporalities of decision making and preventive cyber operations. It suggests that states face a particular challenge in reconciling the different timelines of cyber statecraft practices like military-intelligence operations, diplomacy and foreign policy, leading to institutional tensions and 'spatial-temporal angst'. The last section focuses on the UK and the relationship between the management of geopolitical cyber risk and the construction of political subjectivities inherent in its 'whole-of-society' approach to cyber power. Through the lens of the 'societalization' of risk and resilience, it enrols enterprises and individuals as cyber security actors with nationally relevant roles and responsibilities. This section concludes with a speculative query about the implications of the UK's shift towards a 'war footing' for non-sovereign agents of national cyber security and cyber risk.

Preventive risk management

For states, geopolitical cyber risk is a core feature of national security planning and posture. Since the late 2000s, the UK has adopted an explicitly risk-based approach to national security that addresses 'emerging possibilities, vulnerabilities and preventative strategies' (Neal, 2019: 243). In UK cyber policy, cyber risk is framed as both the risk from known threats, like Advanced Persistent Threat groups and cybercriminals, and from systems vulnerabilities and dependencies, a combination of outwards- and inwards-looking perspectives on the origins and character of cyber risk (Backman and Stevens, 2024: 2451–2). Underpinning both is the appreciation of risk as 'possible outcomes that we can describe in terms of their chances of occurrence, and the impact they would have if realised' (National Cyber Security Centre [NCSC], nd, a). External cyber threats are a striking feature of a 'contested geopolitical environment' (HM Government, 2021: 20) populated by state military and intelligence agencies, cybercriminals and other actors intent on disrupting UK digital assets across the public and private sectors. Government's role is both to stimulate enterprises to manage their cyber risk – through advice, guidance, standards and regulation – and to manage the overall cyber risk to the UK. This latter endeavour requires several lines of inter-related effort.

The first is to manage risk within an overall drive to increase cyber resilience across government digital estate and functions, 'from the delivery of public services through to the operation of national security apparatus' (HM Government, 2022: 13). This necessitates a similar suite of security controls and risk management practices familiar to enterprises, from the 'visibility and understanding of assets held and operated' to threat intelligence, risk assessment and assurance, and governance mechanisms (HM Government, 2022: 29). As with organizations, this enables cyber resilience capacity and business continuity in the provision of essential services but also the ability 'to prosper and, in turn, exercise influence beyond its borders' (HM Government, 2022: 13). Resilience enables organizations to operate transnationally too, but state risk and resilience practices can incorporate a second approach, drawing upon tools and techniques not available to businesses.

Whereas firms cannot usually conduct legal 'hack back' against specific cyber threats (Broeders, 2021), no such strictures prevent states from doing so if they abide by international law and norms of responsible state behaviour in cyberspace (Moynihan, 2021). Cyber operations can be used to disrupt, deny, degrade or destroy the digital assets of state and non-state targets (Moore, 2022) and are often deemed a more palatable alternative to kinetic operations (Smeets, 2018). The UK signals its willingness to use sovereign cyber capabilities against military and non-military threats that 'put at risk our own ability to benefit from a free, open, peaceful and secure cyberspace' (National Cyber Force, 2023: 4). This proactive approach to neutralizing or disrupting threats also underpins the US 'persistent engagement' posture that demotes the utility of cyber deterrence in favour of directly contesting the operational domain and thereby shaping the norms of 'agreed competition' in cyberspace (Fischerkeller et al, 2023). Both US and UK cyber postures can be interpreted as forms of 'preventive risk management' (Kaminska, 2021), although neither fits into standard risk management frameworks because of their incorporation of 'threats' into the *risk = likelihood x impact* equation. This aspect of national risk management seeks to mitigate threats as a way of reducing the cyber risks created by the potential exploitation of vulnerabilities and consequential harm (Heng, 2018). At present, there is little evidence that this approach is effective, nor precisely what capabilities or metrics would be necessary to assess its efficacy (Maschmeyer, 2024).

Another innovation is the mainstreaming of cyber threat intelligence (CTI) in organizational decision making. CTI borrows from military and intelligence methodologies for producing knowledge and understanding for 'decision support', in particular the well-known (and much-critiqued) 'intelligence cycle' (Ainslie et al, 2023). CTI's principal purpose is to assist decision making regarding 'the complex trade-offs involved in addressing, deferring, and accepting risks across varying degrees of uncertainty driven by incomplete, conflicting, and deceptive information' (Work, 2020: 282; Slayton and Muller, 2025). Commercial and government CTI provide indicators of activities driven by (geo)political and criminal motivations and operating across national borders, enabling mitigation of the risks thus perceived to the organizations in question. This is useful for public and private organizations likely to be targeted by cyber espionage actors and for critical infrastructure operators that may attract networked forms of sabotage and extortion. It helps to distinguish between state and non-state actors too, in that different risk management measures can be brought to bear on specific sets of problems. The risk of breach by a Chinese commercial espionage APT will be relevant to some organizations; others will be more interested in their exposure to cybercrime risk originating in Russia or Eastern Europe. Contextualizing an organization's exposure to risk derived from geopolitical competition implies a much broader appreciation of risk than 'cyber' alone, and organizations with the appropriate wherewithal will avail themselves of other services provided by the wider security community.

Against the backdrop of a steady increase in geopolitical cyber activity, there are also 'spikes' in scale and intensity that correspond to specific events or processes in international affairs. Iranian hackers responded to the US bombing of Iran's nuclear facilities in mid-2025 by targeting US banks, defence contractors and energy companies (Klepper, 2025), and other international clashes are always accompanied by inter-state cyber engagements (Maness et al, 2023). Russia's war in Ukraine has been associated with a surge in geopolitically motivated Russian cyber operations against European and North Atlantic Treaty Organization (NATO) targets (Willett, 2024: 118–19). Many of these operations were perpetrated against critical infrastructures mostly owned and operated by the private sector, meaning that companies and governments have been forced to adjust in tandem their risk assessment and risk management positions. For appropriately situated and willing governments – risk appetites vary according to political context and strategic culture (Devanny, 2022; Kostyuk et al, 2025) – the conflict has prompted use of counter-cyber measures and other covert means of preventive risk management to degrade Russian capabilities against Ukraine and its allies, although these have been primarily defensive to date (Kostyuk and Brantly, 2022). The war in Ukraine and wider geopolitical uncertainty have been linked to greater uptake of cyber insurance and claims made, and to increased cyber insurance premiums and the normalization of 'war exclusions' in cyber insurance policies, as underwriters struggle to balance coverage with profit (Lee, 2023). It has also heightened awareness of the need for resilience in the face of geopolitical and systemic cyber risk (Farrand et al, 2024). Managing the sources of geopolitical cyber risk affects all actors in the cyber security ecosystem and drives change in multiple aspects of CRM practice.

Reterritorializing a global domain

The riskification of national security incorporates a greater emphasis on threats than we find in enterprise CRM. In theory, states are securitizing actors that identify issues as security threats and mobilize exceptional responses that may circumvent democratic process and oversight (Buzan et al, 1998).

In practice, state actions in response to threats are often not couched in existential or extraordinary language but align instead with the mundane management of threats and risks, which allows for the movement of issues along a continuum of risk and threat (Huysmans, 2004, 2011; Von Lucke et al, 2014). This applies to state cyber risk to the extent that 'plural risk rationalities' are at work within state-level CRM, including a modality of 'risk as potential threats' that complements a more conventional approach that views risk through the lens of uncertainty (Backman and Stevens, 2024). This enrols known, or 'given', cyber threats (HM Government, 2021: 126) within the wider articulation of cyber risk, which also incorporates less well-known systemic and distributed vulnerabilities in digital networks and systems. The latter are central to national and organizational resilience, the implications of which were discussed in Chapter Six, but the consolidation of threats within national CRM creates new spatial configurations through the deployment of national capabilities and powers.

In proactive and preventive CRM, in which foreign cyber threats are actively identified, pursued and countered through cyber means, there has been a shift in some countries, like the US and UK, to 'persistent engagement' with adversaries in digital networks (Schneider, 2019). Central to this concept of operations is the construction of the global 'network of networks' as an operational 'domain', originally by NATO and now a standard feature of national cyber strategies (Branch, 2020). This unifying move creates a universal operating environment that facilitates greater freedom of movement and sanctions the prosecution of military, intelligence and law enforcement cyber operations outside national borders. While these have always occurred to a certain extent, such as for intelligence collection (Smeets, 2020: 446), in the case of US Cyber Command this doctrinal innovation means that it can operate militarily 'everywhere, all the time and in all ways' (Schneider, 2019), consistent with the US political-military imaginary of unfettered global reach and power projection

(Steger, 2008: 234–41). The 'global', however, is not an unstructured or abstract spatial unit, as it is compartmentalized in important ways.

Reterritorialization of the global cyberspace domain is achieved through the construction of blue, red and grey 'cyberspace areas' (US Joint Chiefs of Staff, 2018: I-4–5). 'Blue' spaces are those for which a government or other 'friendly' entity has direct responsibility, owns, controls or is otherwise directed to protect. 'Red' spaces are owned or controlled by an adversary. Spatial zones that do not qualify as either blue or red are termed 'grey' spaces (Smeets, 2020: 445–6). This terminology mirrors standard nomenclature in network security exercises, often conducted as part of organizational risk assessments, in which enemy 'red teams' attempt to access and subvert the friendly networks of 'blue teams' in order to ascertain technical and processual vulnerabilities for mitigation (Jabbour and Poisson, 2016). Blue, grey and red spaces do not map to conventional territories, even though they may overlay particular polities and topographies. They are topological constructs subject to dynamic change and deformations in networked space, as actors win and cede control over digital assets through constant competition. The configurations of blue, grey and red spaces are contingent on prevailing behaviours and can only be temporary stabilizations under conditions of persistent engagement. Cyberspace operations as part of geopolitical risk management are therefore complicit in the co-production of the spaces of the state and its adversaries. Blue space is stretched to wherever friendly forces operate or where national digital assets are situated, such that cyber operations like 'hunting forward', 'active defence' and 'hacking back' reconfigure the boundaries of the state and complicate territorial notions of its 'inside' and 'outside' (Lambach, 2020; Branch, 2024). These topological deformations are operational in origin but also translate into normative claims about what comprises the territory of the state in cyberspace, with states asserting sovereignty over 'their' networks, data and citizens located in the physical territories of other countries (Zeng et al, 2017; Santaniello, 2025). If red space increases as cyber operations expand, or blue space is claimed abroad, tensions will inevitably arise between nations, allies included, when these new spatialities conflict with established territorial constructs (Smeets, 2020: 446).

The incorporation of threats into the risk calculus does not do away with territorial binaries altogether. The language of inside/outside continues to operate as a powerful demarcation between the 'core' nation to be protected and assailing forces originating along and beyond the perimeter (Clapton, 2014). Both US and Canada policy name specific countries – China, Russia, Iran, North Korea and, in Canada's case, India – as sources of geopolitical cyber threat and risk (White House, 2023; Canadian Centre for Cyber Security, 2024). Others are more circumspect, with the UK, for example, usually refraining from identifying countries by name in policy, referring instead to more nebulous categories of 'state or non-state groups' or 'state-sponsored hostile activity' (HM Government, 2021). Public attributions of distinct cyber operations to foreign governments are, in contrast, increasingly common (Baram and Peer, 2024), with states previously reluctant to name particular countries, like France, also shifting to a more robust position on 'naming and shaming' their strategic adversaries (Jalabert et al, 2025). In all cases, though, risks emerge from outside the home polity, a 'spatialization of good and evil' in a moral cartography that is at variance with the transversal reality of sociotechnical cyberspace (Bialasiewicz et al, 2007: 417–18).

The construction of spaces external to the state as threatening or risky is commonplace in cyber security (Barnard-Wills and Ashenden, 2012) and serves clear functions in security politics, but it also constrains 'internal' actors in unusual ways. There has, for instance, been a historical reluctance on the part of Western CTI companies and media organizations to report on the offensive cyber operations of Western countries, neither wishing to disturb status quo relationships with their home governments (Martin et al, 2024; Slayton and Muller, 2025). Research indicates that these considerations are becoming more acute, as geopolitical calculations affect commercial decisions about whether to expose Western cyber operations, resulting in less transparency than ever (Yoffe et al, 2025). This taciturnity has led to non-Western powers like China capitalizing on the situation through strategic communication exposing alleged Western hypocrisy in attribution practices (Iasiello, 2024). It has also spurred non-Western countries to accelerate their own capacities to attribute cyber operations to Western nations (Hurel, 2025). In seeking to manage geopolitical cyber risks to the 'inside' of the territorial binary, actors inadvertently create other problems that are reflexively also constructed as risks.

Precaution and angst

The issue of unintended or undesirable consequences influences how states manage cyber risk. Epistemic uncertainty about the character and knowability of future cyber threats translates into risk aversion as a dominant factor in national decision making, despite the propensity of national security officials to underestimate cognitive uncertainty (Friedman, 2025). Monica Kaminska (2021) argues that Western societies are noticeably reluctant to respond robustly to known cyber threats because they fear the possibilities of operational blowback, escalation and disclosure of their sovereign capabilities. This is consistent with a reading of Western societies as 'risk societies', which are reflexively concerned with the risks generated by their own actions. Geopolitical CRM is an attempt to forestall future negative outcomes, as would be expected, but also works to constrain or defer one's own actions for fear of making a problem worse, a defining characteristic of cyber security as a 'wicked problem' (Carr and Lesniewska, 2020). This does not automatically imply inaction in the present, but it shifts the emphasis of policy to 'non-obvious' (Libicki, 2012) preventive measures targeted at specific cyber threats, and to cyber resilience and incident management that accept the need to deal with future materializations of cyber risks (Backman and Stevens, 2024).

This is the 'precautionary principle' at work, which states that a lack of certainty about future harms should not overly inhibit remedial actions in the present (McLean et al, 2009; Ahteensuu and Sandin, 2012). This principle – or approach – does not dictate whether specific actions are allowed or prohibited, only that uncertainty about the future should not be a hard constraint on policy actions in the present if they are likely to produce positive future outcomes. In cyber security, the future is always uncertain, as are the possible countervailing risks (Graham and Wiener, 2009) of one's own actions, so the precautionary approach enables actions that are also 'risky' but which may alleviate the emergence of future threats. Calculations about precisely what to do and how are a function of risk appetite and of 'risk–risk' trade-offs. While this can lead to apparent inaction in response to cyber threats and risks (Kaminska, 2021), neither a diminished appetite for risk nor the need to trade off risks against one another is a bar to the use of state power (Hansen et al, 2008). Preventive risk management through state cyber operations is situated in this register as a set of techniques for reducing risks by countering and neutralizing cyber threats before they can harm the referent object of the nation.

Studies of the precautionary principle in counterterrorism indicate the emergence of risk as 'a modality of governing and ordering reality' (Aradau and Van Munster, 2007: 108). Like terrorism, cyber risk is inherently uncertain and therefore incalculable; unlike terrorism, however, the risks of which are contingent upon imaginations of future irreversible catastrophes, cyber security has moved beyond its discursive and political reliance on 'cyber doom' (Lawson, 2021) to motivate state policy and practice. The catastrophized emergency still forms part of overall security and risk anticipation, such as in the UK's categorization of cyber incidents (NCSC, 2023d), but it has become much less pronounced as governments recognize that the cyber threat is constant rather than reducible to a series of large-scale events (Farrand et al, 2020). Decision making needs to address future possibilities, particularly hostile state cyber activities or ransomware operations against critical infrastructure, but it recognizes the 'campaigning' nature of adversarial operations and of state countermeasures (Harknett and Smeets, 2022). This implies an extended temporality of decision making and action that is effectively infinite, and a diversity of operational types and tempos within this temporal frame. As if to illustrate the co-construction of the preventive risk temporality and political action, the former head of US Cyber Command identified the need to ensure 'our capabilities, operational tempo, decisionmaking processes, and authorities enable continuous, persistent operations' (Nakasone, 2019: 12). The extended temporality of prevention demands appropriate material, cognitive and political resources, which in turn enable that same temporality of operational engagement.

The frenetic operational tempo of cyberspace means that decision making often occurs 'under the burden of compressed decision timelines', which creates 'spatial-temporal angst' in the minds of decision makers (Cunningham and Tomes, 2004: 134). This condition is constituted by 'increased awareness of distributed (and global) effects of decisions, uncertainty about interaction effects of such decisions, and a leadership disposition against risk-taking' (Cunningham and Tomes, 2004: 134), all of which we can discern in geopolitical CRM. The temptation in such a situation is to accelerate one's own operations and adoption of proactive measures, as outlined explicitly by one of the architects of US persistent engagement (Harknett, 2020). This helps align the operational tempos of preventive CRM with the temporalities of the environment and of hostile actors but does not alleviate the wider epistemic and political problems of decision making, uncertainty and risk perception.

Managing cyber risk at the state level entails other non-military forms of 'cyber statecraft' (Healey, 2011), including foreign and economic policy, diplomacy and productive engagement with other important actors, including international organizations, civil society and the private sector. Each operates in temporal frames that are difficult to align in national cyber statecraft and CRM or through an overarching narrative of 'timing' that imposes order on multiple temporalities (Hom, 2020). The tempo and duration of military-intelligence cyber campaigns differ substantially from the longer timelines of diplomacy intended to broker international agreements on cyber risk and threat reduction, for example. The drafting, publication and implementation of cyber strategies with risk management at their core will inevitably appear too drawn out and attenuated to address a cyber threat that is 'severe and advancing quickly' (National Audit Office, 2025: 13). The fundamental tension in CRM between short-term reactivity and long-term planning and strategy is exacerbated by the state's need to integrate multiple time horizons, actors and areas of responsibility.

Societalization of cyber risk

Just as the state approach to CRM stretches the conceptual apparatus of risk through the reincorporation of threats (Backman and Stevens, 2024), it also enrols new actors and agents into its strategic framework. This moves beyond a now conventional appeal to the private sector to cooperate fully with governments to secure critical infrastructure, incentivize cyber security as a career path or develop products that are 'secure by design'. Instead, emphasis lies in developing 'whole-of-society' approaches that seek to understand enduring security problems and harness the skills and expertise of all stakeholders, from industry to academia, government to civil society. As cyber security and cyber risk are problems for all parts of society, there too must be found the solutions: 'what happens in the boardroom or the classroom matters as much to our national cyber power as the actions of technical experts and government officials' (HM Government, 2021: 11). In the UK, this link between its 'whole-of-society approach' and 'cyber power' is explicitly recognized as leveraging all the resources and opportunities of the nation in order to develop 'the ability of a state to protect and promote its interests in and through cyberspace' (HM Government, 2021: 20). This connection is not conceptually novel (see for example Klimburg, 2011), but it has taken on marked policy salience in recent years. The generation of cyber power has appreciable implications for foreign policy, diplomacy and cyber operations (Barrinha and Renard, 2020: Stevens and Kavanagh, 2021), but its domestic governmental effects have rarely been addressed.

The role of cyber security as an economic driver is a common feature of cyber strategies, and governments have positioned themselves as enablers of markets for cyber security innovation and growth (Cobos et al, 2024). Over time, successive governments have recognized that the market has struggled to deliver either outcome, necessitating a slightly different role for the state and for the private sector. The latest US national cyber security strategy states: '[W]e must shape market forces to place responsibility on those within our digital ecosystem that are best positioned to reduce risk' (White House, 2023: 19), that is, those firms that own and operate most digital infrastructures. The intention is not to assert state power through direct intervention in the market but to develop and encourage 'practices that enhance the security and resilience of our digital ecosystem while preserving innovation and competition' (White House, 2023: 19). The strategy is also keen to insist on the accountability of firms for their activities and, overall, stresses their essential role in contributing to successful national CRM. The explicit responsibilization of this reframing of cyber risk and resilience manoeuvres firms into new subject positions within the political economy of cyber security and cyber risk (Weiss and Krieger, 2025). For some companies, this is an opportunity to orient themselves as trusted partners and vendors (Mathur, 2025); for others, it occasions new costs and constraints in the form of regulation, legal liabilities and increased visibility of – and accountability for – their actions.

The redistribution of responsibility and accountability from the state to sub-state entities like firms, communities and individuals has been described as the 'societalization' of security (Chandler, 2013, 2018b), in which the focus on the state as the primary producer of security and its allied practices shifts to society as a whole. This is a salient feature of cyber resilience, where surviving and thriving in the context of cyber insecurity and cyber risk is enshrined in policy as the responsibility of actors from the top to the bottom of the socioeconomic hierarchy and in practice enrols stakeholders 'at the edge' of the cyber resilience assemblage (Holmström and Große, 2025). Couched as empowerment in response to uncertainty, the teleological nature of cyber resilience – accidents *will* happen – means that there always remains the lingering suspicion that either the state cannot protect 'resilient subjects' from cyber threats or that it will not. No government will articulate either position openly, of course, even if they did believe this to be the case, preferring instead to stress the need to enrol all actors in the management of cyber risk as a common-sense response to a set of problems that demonstrably do affect all members of society, albeit differentially. As identified in Chapter Six, however, there are limits to the agency and adaptability of the subjects of cyber resilience and open questions about the political implications of resilience as a form of governmentality.

In the context of geopolitical cyber risk, societalization constructs the subjects of risk and resilience in an additionally problematic light. The whole-of-society approach emerged as a policy response to the perceived need to harness the skills and expertise of all sectors of society while energizing a collective sense of the national cyber security mission (Harknett and Stever, 2009). In addition to engaging the private sector, governments have also extended their efforts into local communities, including through school awareness and education programmes. In respect of skills development and the establishment of career paths for future cyber security professionals, this addresses an important capacity issue while also raising uncomfortable ethical questions about the construction of children as 'agents' of security (Stevens, 2016: 172–6). Occasioned by both the actions of adversaries (principally, Russia) and allies (United States), recent UK policy has sharpened the importance of the whole-of-society approach to wider national security posture and objectives (HM Government, 2025c). Policy allusions to 'war readiness' and putting Britain on a war footing are accompanied by UK households being told to 'prepare for war', a reframing of the UK's strategic posture and the roles that British citizens are required to adopt in this new geopolitical context (HM Government, 2025c; Whannel, 2025). Irrespective of the empirical validity of this proposition, or of its operational necessity should worst-case scenarios transpire, the background transformation of cyber security into a pre-war project to counter geopolitical adversaries implicitly militarizes civilian actors and the operational environment. No longer 'just' security actors, civilian contributors to cyber security, CRM and cyber resilience are reconstituted as essential actors in a nation either already at war or soon to be in one.

Conclusion

Government approaches to geopolitical cyber risk encompass a wider range of risk variables, practices and objectives than those of individual organizations, constructing a category of CRM that is highly contingent yet shaped profoundly by strategic imperatives. It is informed by qualitative judgements more strongly than organizational forms of CRM. Governments differ too in the scale and purview of their responsibilities, the national and economic security stakes of getting it right or wrong and in their explicit anxiety over known cyber threats. In the latter register, the state management of cyber risk corresponds with more conventional threat-based national cyber security. Governments continue to address risk as uncertainty while also developing new forms of preventive CRM that tackle risk as possible threats. These involve proactive measures beyond the territorial boundaries of the state, seeking out and neutralizing threats that pose risks to the nation, and in so doing create new spaces for exploitation and manoeuvre. The complexities of geopolitics mean that cyber statecraft also involves a range of practices beyond military and intelligence affairs, with different timelines and decision-making pressures. In attempting to develop the appropriate resources for generating effective international action against the backdrop of increasing geopolitical competition and war, the whole-of-society frame has emerged to inspire and harness the innovative and adaptive capacities of companies and citizens. This prompts us to query the relationship between state and sub-state entities, whose capacities, responsibilities and subject positions are reconfigured not just through the everyday needs of cyber security and CRM but relative to geopolitical context and political rhetoric.

This final observation returns us to governmentality, in its concerns with how human behaviours are influenced and regulated through dispersed rather than sovereign techniques and technologies alone. As the four chapters on CRM forms have shown, governments play decisive roles in establishing frameworks for enterprise CRM, cyber insurance, cyber resilience and geopolitical cyber risk, but the emphasis is very often on creating the conditions necessary for non- or sub-state actors to take active roles in the management of cyber risk. Chapter Eight concludes the book by returning to the main theme of governmentality and exploring some of its implications and futures.

EIGHT

Conclusion

The previous chapters have described the logics, practices and effects of four main fields of cyber risk management (CRM). Each modality of CRM differs from the others in significant ways, yet they articulate with one another too. One of the tasks of enterprise CRM, for example, is to decide how to manage residual risk, which is where cyber insurance plays a crucial role. Cyber risk managers will also invest in cyber resilience, either as a specific set of practices and objectives or, increasingly, as an overarching rubric for all their cyber security and cyber risk investments. Cyber resilience depends on good cyber security and effective CRM, including the use of cyber insurance to defray the costs incurred when security and risk management fail. CRM, insurance and resilience are situated along a continuum of security–risk–resilience, as in geopolitical CRM, where states draw upon resources from all points on this spectrum to manage cyber risks emerging from beyond their borders. These include preventive measures like out-of-area cyber operations that do not align easily with traditional perspectives on risk, but which serve a risk management function in theory and in practice. The overall field of CRM is clearly evolving beyond any classically 'pure' vision of risk analysis and assessment and into a realm of rationality and practice that is vibrant, questing and experimental in response to the actual and perceived uncertainty of the digital world.

While respecting the heterogeneity of CRM, it is necessary to offer some concluding thoughts on the governmentality of CRM in the round. A more detailed account of this problem is a task for future research, but we can provide preliminary support for the thesis that CRM expresses governmental logics discernible through analysis of its discourses and practices. This concluding chapter is arranged in three sections. The first returns to the principal theoretical lens of the book, governmentality, and offers a synthetic treatment of the cyber risk *dispositif* viewed in these terms. It reasserts the multiplicity, relationality and sociotechnical character of CRM and the topological nature of power in this field of policy and practice. The second section identifies the implications of the governmentality of CRM, arguing that while it exhibits a lack of formal political contestation, it is still political. The construction of hazards and dangers as risks is a political choice, as are the measures developed to manage those risks. This section argues that three 'differences' are important to understanding the governmentality of cyber risk: perspective, responsibility and the shift to preventive risk management. The third section outlines a selection of possible futures for CRM. It discusses the continuation of CRM under conditions of digital transformation, its enrolment of artificially intelligent agents and the significance of risk–risk interactions. The chapter concludes by reiterating the main objectives of the book and the opportunities it provides for further analysis.

Governmentality

The cornerstone of CRM in all its forms is the reduction of risk to an entity from future events that will cause it harm through digital means. Its focus is primarily on the referent object of security – the individual, the organization, the nation – over and through which governmentality operates. In contrast to cyber security – although risk and security are interrelated in theory and practice (Weiss and Jankauskas, 2019) – risk is less oriented to countering known threats than with mitigating the impact of future, as-yet-unknown threats, the identities of which are inevitably unknowable in their totalities. CRM in all its guises is an admixture of science and speculation, quantitative and qualitative methodologies, experience and intuition, that combine to 'see things in their process of emergence' (Chandler, 2018a: 22). Cyber risk knowledge is constructed through and in direct conversation with perceptions of uncertainty, whether of the operational environment of cyberspace or of the unknowability of future threats and risks. The identification of a cyber risk *dispositif* heeds this epistemic and ontological uncertainty. It exists in a state of permanent flux, its boundaries uncertain and mobile, its drivers and effects contingent on multiple economic, social, political and technological factors (Whiting, 2020: 51). Yet it retains enough internal coherence as a temporary stabilization of affairs to be a useful unit of analysis, within which we can identify provisionally the dynamics of governmentality.

As a means of shaping the 'conduct of conduct', CRM operates across diverse areas of society, enrolling humans and nonhumans in fields of discourse and practice whose influence is exerted often quietly but always with governmental effects. Cyber risks are brought into being as objects to be governed by the strategies and practices of CRM. Its techniques include policies, regulation, standards, guidance and training that shape the conditions of possibility for the emergence of risk and of the subjects of cyber risk, whose agency is constrained and enabled in manifold ways. These are underpinned by specific technologies of risk – risk assessment and analysis, security controls, insurance underwriting, resilience frameworks, cyber operations – that are fundamentally sociotechnical assemblages of people, organizations, data, devices and software. These configurations are vigorous and relational, arranged strategically and in combination to bring about new forms of spatiotemporal ordering that facilitate the management of cyber risk and its exploitation.

The operations of power in the cyber risk *dispositif* are both sovereign and non-sovereign. The state provides significant momentum to the practices described in this book, but it does not operate in isolation. Many innovations in risk thinking and practice derive from non-state actors like insurance and CRM companies, or from international standards bodies, who see opportunities for meaningful intervention (Prakash and Potoski, 2010) and, in the case of firms, economic gain. These 'entrepreneurial' or 'protean' powers are 'the effect of improvisational and innovative responses to uncertainty that arise from actors' creativity and agility in response to uncertainty' (Katzenstein and Seybert, 2018: 4). States provide the legal, regulatory and often normative frameworks that circumscribe aspects of these activities, but they are conducted through global networks of capital and digital infrastructure and so exert transversal governmental effects that escape the 'control' power of the state. The circulation of powers in the governmental assemblage is always manifold and impelled by different rationalities and strategic objectives, which sometimes align but not axiomatically so. One actor's attempts to reduce risk can increase risk for another. A state's use of preventive cyber operations, for instance, potentially erodes the cyber security of the targeted state; the risk calculus of each is thereby affected and will feed back into future decision making and behaviour (Brantly, 2016). Similarly, trade-offs are made between actors working across numerous sites and sectors, through which the management of cyber risk becomes a 'heteropolar' site of political contestation and confrontation (Deibert and Rohozinski, 2010). The recent history of digital affairs suggests that the state, for instance, is not retreating from its cardinal role, nor that it is wholly delegating its responsibilities, but that it is reasserting its authority through hybrid forms of authority, influence and governmentality (Haggart et al, 2021). Cyber risk knowledges and practices are malleable and contested in and between communities, and further research would illuminate how the powers of relevant actors converge and diverge in CRM.

The sociotechnical aspect of the cyber risk *dispositif* is central to understanding how power operates in CRM. It emerges not only from centres of state calculation and control but from distributed relational assemblages of agents, institutions and risk technologies, the latter often mobile and 'redeployed' in multiple contexts (O'Grady, 2016). These configurations are no longer territorially bound, leveraging instead the networked topologies of technologies and markets 'to make their presence felt in more or less powerful ways that transcend a landscape of fixed distances and well-defined proximities' (Allen, 2016: 2). This topological reading of power is the inverse of the conventional understanding of power as something that diminishes with distance. Instead, actors have 'reach' within topological assemblages 'to reproduce their power and advantage by directly folding in their influence from afar or by indirectly stretching their authority through time' (Allen, 2016: 82). The deformation of networked topologies allows actors to be relationally 'present' even if geographically or temporally distant, which helps explain how many aspects of CRM are experienced transnationally and transversally and how actors exert governmental effects that are distributed in space and time in complex sociotechnical assemblages. The powers at work in the governmentality of CRM are plural, relational and topological.

Governmentality is already plural, as are the sociotechnical environments in which it generates its effects. It has no single governor or guiding ideology and operates through heterogenous rationalities and modalities that work together towards common goals in hybrid fashion. Elements can be identified with sovereign control or with neoliberal motives, but it cannot be reduced to these without truncating the empirical complexity of CRM, the extent of which is far greater and more nuanced than such parsimony would allow. Investigations of different national or cultural contexts, or specific communities of thought and practice, would provide additional insights and necessary adjustments beyond the admittedly restricted priorities of the current analysis. However, the essential methodological scaffolding of this book and the theoretical resources at its disposal are equally suited to studies of the governmentality of CRM wherever they are situated.

Implications

CRM is rarely a matter of significant public disquiet. It is not the topic of headlines or fevered public commentary, whereas major cyber attacks do become news items and prompt open questioning of what, why and how they happened. Threats real and imagined attract public and media interest, whereas risk management does not. The founding chief executive officer of the National Cyber Security Centre argued of the UK that 'nobody plays politics with cyber', by which he meant that cyber security was 'an important but not controversial subject' (Villegas, 2025). In the context of parliamentary and party politics, this is true, and cyber security is notable among security issues for its relative bipartisanship and consensus. At times, it can appear almost de-politicized or technocratic, failing to precipitate public or political interest even when significant cyber incidents occur (Martin, 2024b). This does not mean, however, that cyber security is not political, or that it has no political effect. Security is always political because it involves decisions about what is identified as a threat, whose security is prioritized and what responses are justified to address those threats. Risk too – as 'second-order' security (Corry, 2012) – is political: what is to be identified as a risk, what is it a risk to, and how should it be managed? Like security, risk is constructed as an object of corporate and public policy by a complex assemblage of public and private actors with varying degrees of authority and legitimacy (Kruck and Weiss, 2025). Cyber risk is undoubtedly one of the 'minor currents in contemporary politics', but it is still political, and the question to ask of all such issues is 'what difference they induce' (Walters, 2012: 149).

The first difference cyber risk incubates is an alteration in perspective. The referent objects of security/risk do not change, but the perspective shifts from the relative severity of cyber threats to the vulnerability of the object itself. The focus on individuals and organizations is fine-grained, but as the aperture widens and expands to the level of the state, this granularity is erased by a generalization of societal vulnerability that, even with good intentions, may lead to a sense of 'permanent insecurity' (Hagmann and Dunn Cavelty, 2012). A cyber threat can usually be dismissed as 'over there' or outside the safe zones of the homeland; cyber risks are always embodied 'over here' and within the state or other entity 'at risk'. Cyber resilience, for instance, through its dependence on the radical immanence of threat, is a potential source of vulnerability and anxiety, although data on public perceptions of cyber risk that would corroborate these theoretical claims are scarce and unreliable (Kostyuk and Wayne, 2021). The nature of CRM is that it must construct risks to sustain its identity and economic foundations – it is an open-ended project that requires the conversion of threats and hazards into risks to perpetuate itself. Identification of risks presupposes the existence of vulnerability and the possibility of harm, so CRM is implicated accordingly in the construction of individual, organizational and societal vulnerability. In an additional perspectival twist, policy makers must also contend with the realization that they are embedded in the very risk objects they seek to govern and in epistemic frameworks from which it is hard to escape: there is no longer any possibility of governing risk 'from the outside' (Chandler, 2018a: 61).

This leads to a second difference induced by CRM and concerns the loci and boundaries of responsibility. As a distributed yet socially constructed phenomenon, cyber risk is both everywhere and nowhere at the same time. 'The' cyber risk is global, but 'a' cyber risk might be local and specific or, very often, not specified as a risk in the first place. The current direction of public policy is to raise awareness of cyber risks, which implies that cyber risks are objective phenomena that can be communicated and understood with minimal translation or semiotic slippage. The task of authoritative parties is to educate and inform the unaware of the existence of cyber risks, so that they can address them in line with higher-level strategic objectives. This is governmentality at work, as actors retain their freedom to choose while being reminded of the penalties for not acting 'correctly', which may include legal and regulatory liabilities or an inability to conduct business with responsible others. The responsibilization of actors is clear in all areas of CRM and is explicit in the 'whole-of-society' approach to cyber risk and cyber security. Many actors embrace these responsibilities as the 'right thing to do' for themselves and society, which again confirms the self-regulatory aspects of governmentality, but it raises questions about those who cannot participate in these activities for lack of cognitive and financial resources, or who refuse the construction of cyber issues as risks. What becomes of them, and who will be responsible for their wellbeing?

It is not possible to draw bright moral lines through these issues: to embrace or refuse cyber risk is a choice based on situated ethics, norms, resources and values. The third difference also displays this tension: the adoption of risk management as a framework for state action in cyberspace feels both necessary and unfortunate. Preventive CRM through cyber operations is seen as a crucial means of countering cyber threats before they effect harm on vulnerable infrastructures and populations. Risk-averse states use these tools sparingly; states with greater risk appetites are bolder and more profligate. The actions of the latter force the hands of the former, as they pose risks that cannot, politically or practically, be left unaddressed. This leads to the development of doctrines that enable out-of-area peacetime operations (all bets are off in war) that may in turn generate new risks of reputational damage, public scepticism and diplomatic complications, not to mention the operational responses of strategic adversaries or the unintended consequences of one's own actions. There is no straightforward escape from this escalatory dilemma. The incorporation of direct engagement with cyber threats into the risk management frameworks undoubtedly complicates how we conceptualize national CRM. The best governments can do is conduct preventive operations responsibly and with the unequivocal minimization of civilian harm while explaining their operational and strategic posture to their citizens and external parties. Concurrently, they must accept that preventive action does not 'solve' the overall problem of hostile cyber activity.

Futures

There are conceptual limits to the riskification of cyber security: not everything will be constructed as a risk, nor can every issue be addressed through the tools of CRM. However, as the threat environment evolves and digital transformation proceeds apace under conditions of uncertainty, there are no practical limits to the continuation of risk management as a mode of governing the digital and its dependencies. New threats will arise to pose new risks to entities large and small. New vulnerabilities will be discovered in existing sociotechnical systems, and every system brought online will bring its own problems and peculiarities that will occasion risk analysis and mitigation. In response, the CRM industry will develop novel tools and approaches for managing cyber risk; cyber insurance will become more sophisticated; and cyber resilience will sustain genuine adaptation as cyber risks materialize in perpetuity. CRM may become so effective – and cyber security so good – that an equilibrium is reached. At that point, we can reap the rewards of digital stability hard-won from decades of effort and experience. Somehow, this seems unlikely, and the track record of risk management as a means of addressing large-scale problems affecting humanity is less than impressive (Dalby, 2013). Whichever imagined future we prefer suggests that CRM will operate long into the future as a means of attempting to manage vulnerability and harm, regardless of its success or otherwise.

From the perspective of governmentality, whether there will be more or less risk in the future is not as important as how risk technologies will be redistributed, reshaped and 'valorised in new ways and forms' (O'Malley, 2004: 174). The governmentality of cyber risk will persist and deepen, finding new modes of expression and influence and new conduits for the microphysics of power in sociotechnical assemblages. The prefix 'cyber' offers something of a distraction in thinking about this future, as it creates the impression of a 'place apart', as early theorists of cyberspace were keen to argue (Cohen, 2007; Lambach, 2020). As information technologies become embedded in all aspects of human activity, environments and bodies, the construction of cyberspace as separate to 'real' life is increasingly unsustainable. Information technologies do create virtual spaces, but the technologies themselves, in their material and infrastructural dimensions, are as much the fabric of the world as any other technology. More so, perhaps, as rapid digital transformation adds billions of new nodes of data creation and exchange, interconnected through the protocols and machinery of digital networks. The dynamics of globalization, in which the distribution and embeddedness of digital technologies are both driver and effect, intensifies the reach of governmental actors and binds them ever more fundamentally into fields of transversal action (Larner and Walters, 2004). The 'attack surface' of cyber insecurity is extended across the global information environment while becoming imbricated in the built and embodied architectures of daily life. It is less a 'surface' than a three-dimensional sociotechnical matrix of vulnerability and potential harm and therefore emergent cyber risks.

The technologies of CRM will mirror these developments, seeking out new opportunities for shaping the 'conduct of conduct' in increasingly fractal digital assemblages. Driven by motives of protecting populations from harm, securing information systems and corporate gain, cyber risk governmentality will intensify in reach and function, aiming not to control but to enable the flourishing of technological life in all its social and economic dimensions. More accurately, cyber risk managers will modulate the cadences and content of digital activity congruent with their strategic objectives, manipulating forms of power and technique in the cyber risk *dispositif*. More of us will become agents in this endeavour, either charged with specific risk management tasks in the workplace, or de facto responsibilized through the policies and practices of security, risk and resilience. The 'whole-of-society' framework means just that: everyone must play a role in addressing and managing cyber risk.

We are being joined by a new class of actors in cyber risk, as we are in many areas of life: artificial intelligence (AI). This book has alluded several times to the enrolment of AI into the cyber risk assemblage, and its contribution to the automation of anomaly detection, cyber threat intelligence, risk assessment and incident response. AI empowers attackers with new tools, facilitates cyber threats at scale, automates vulnerability discovery and exploitation and, through the widespread deployment of poorly understood and weakly secured AI technologies, creates new vulnerabilities that must be identified and mitigated (Bonfanti, 2022). In the attempt to use AI to reduce uncertainty about cyber risks and threats, AI introduces new sources of uncertainty – about its efficacy, the accuracy of its truth claims and decision making and the pernicious effects of its actions. Although these problems are well known, they are unlikely to seriously constrain AI deployment, including for national cyber security purposes, which in turn generates new forms of risk (Jensen et al, 2020; Whyte, 2023). The presence of AI and algorithms as co-producers or even primary producers of cyber risk knowledge also suggests the operations of 'algorithmic governmentality' (Aradau, 2023) that short-circuits human decision making and interactions between human subjects in the cyber risk *dispositif*. AI will generate additional relations of power and shape the 'conduct of conduct' of risk professionals and end-users in ways not readily discernible to the average observer, if at all.

AI entities are already enablers of CRM, and their presence will become more emphatic and consequential as organizations look to deploy AI wholesale for productivity and efficiency gains. Through its relational agency and autonomy, AI will co-construct the social reality of CRM (Airoldi, 2022).

The intersection of cyber risks with AI risks adds an additional dimension to the risk–risk interactions already identified. As has been noted, cyber risk is not a pristine, isolated risk but converges with and amplifies other forms of risk, including enterprise and (geo)political risk. Risk–risk interactions have the potential for effects in one domain to trigger effects in another. CRM already takes account of, for example, the relationship between supply-chain risks and financial and reputational risks, or the possibility that cyber risks could concatenate and form systemic risks with wider socioeconomic impact. The adoption of any technology contains within it the risk of failure, accident and exploitation (Perrow, 1999), and the convergence of two suites of technology (cyber, AI) that are demonstrably insecure and uncertain presents new challenges for the risk managers of tomorrow. Importantly, both are *information* technologies, feeding off and back into one another in tight feedback loops sustained by the exchange of data written in binary, digital form. There are cyber risks to AI, and AI risks to cyber security, twin vectors for deliberate abuse, accidents of code or faulty inputs and outputs leading to mutually reinforcing, even runaway, negative effects. As society becomes increasingly digital, this informational substrate is the platform upon which all other risks converge. Cyber risk holds not just the potential to be 'a' risk to society but 'the' risk to everything built upon digital infrastructures.

***

CRM fulfils important societal functions that would suffer without the ability to reduce risks to digital systems and end-users. The riskification of cyber security illustrates that the threats and hazards of digital transformation are constructed as risks so they can be anticipated and managed to minimize future harms. This is an uncontroversial proposition and relegates CRM to an issue barely worth discussing, much less contesting politically. Any critical analyst working on risk will quickly recognize that 'there is very little room for counterargument against orthodox thinking' (Deuchars, 2004: 134). The orthodoxy of CRM is that it cannot be challenged – the only contestation comes from within and resolves to arguments about emphasis and methodology, or its relative weighting in policy. It is certainly mostly banal, mundane and, to a considerable extent, technocratic, but this does not mean it is apolitical (Esmark, 2020).

The purpose of this book is to investigate the circulation of powers in the cyber risk assemblage, how rationalities and materialities interact and to introduce insights from governmentality to the analysis of CRM in its various forms. CRM requires the agency and energy of powers beyond the sovereign. Firms are indispensable actors in the cyber risk *dispositif*, as are citizens and civil society, international organizations and the academy. Governmentality is not hegemonic, as it requires the delegation and distribution of sovereign and non-sovereign powers acting transversally and in cross-pollination to achieve higher-level societal ambitions. It is also internally heterogeneous, reflecting the priorities of particular actors, whether these are individuals, organizations, states or international actors (Joseph, 2021). In the case of the UK examined in this book, the management of cyber risk and cyber security is framed politically as a 'whole-of-society' imperative: cyber risk affects everyone, so everyone should be engaged in managing it, albeit in contextual fashion and with varying degrees of responsibility. This is a legible first political move in this specific case, but an analytical move must follow: what are the tangible political effects of CRM?

The objective of this book is not to moralize about whether CRM is somehow 'right' or 'wrong' but to ask questions about how governmentality operates through CRM and to begin to identify those effects. This book has provided some indications of these, but its analysis is provisional and, in keeping with one of its main recurring themes, uncertain. Ulrich Beck (2009: 9), one of the grand sociological theorists of risk, wrote that risk 'signifies the controversial reality of the possible'. This conceptual and methodological proposition is an apt conclusion to the book, as it offers avenues for exploration and critique by investigators of digital politics, risk and security. Researchers across sociology, geography, security studies, international relations, organization studies and many others might pull on any of the threads presented here and conduct more sophisticated and detailed analyses of how CRM is spoken about and practised, the motives and rationalities involved and more precise interrogations of the workings of governmentality and its political and socioeconomic structuring effects. Attention could also be directed more closely to the forms of exclusion and subjectification thus enacted and to the possibilities of resistance, reconfiguration and refusal. Practitioners too might reflect upon their assumptions and actions, with a view perhaps to considering how CRM can be implemented ethically as well as effectively. For policy makers, this book may not provide answers to the wicked problems of digital transformation that they face daily, but it poses questions about the character of public–private interactions and of the contract between states and citizens. As argued throughout this book, the transversal governmentality of CRM shapes the conditions of possibility for social action. Even in its most technocratic manifestations, it is profoundly political.

References

Aanonsen, C.E. and Knudsen, R.A. (2025) 'Riskification and the production of threat: a comparison of risk assessments in cybersecurity and counter-terrorism', *Critical Studies on Security*, Available from: https://doi.org/10.1080/21624887.2025.2509343

Adler, E. and Pouliot, V. (2011) 'International practices', *International Theory*, 3(1): 1–36.

Adriko, R. and Nurse, J.R.C. (2024) 'Cybersecurity, cyber insurance and small-to-medium-sized enterprises: a systematic review', *Information and Computer Security*, 32(5): 691–710.

Adriko, R. and Nurse, J.R.C. (2025) 'Cyber insurance is no silver bullet for cybersecurity', Binding Hook, 10 February, Available from: https://bindinghook.com/articles-hooked-on-trends/cyber-insurance-is-no-silver-bullet-for-cybersecurity/

Ahteensuu, M. and Sandin, P. (2012) 'The precautionary principle', in S. Roeser, R. Hillerbrand, P. Sandin and M. Peterson (eds) *Handbook of Risk Theory*, Springer, pp 961–78.

Ainslie, S., Thompson, D., Maynard, S. and Ahmad, A. (2023) 'Cyber-threat intelligence for security decision-making: a review and research agenda for practice', *Computers and Security*, 132: 103352.

Airoldi, M. (2022) *Machine Habitus: Toward a Sociology of Algorithms*, Polity.

Aitel, D. and D. Geer (2025) 'AI and secure code generation', Lawfare, 27 June, Available from: https://www.lawfaremedia.org/article/ai-and-secure-code-generation

Allen, J. (2016) *Topologies of Power: Beyond Territory and Networks*, Routledge.

Amir, S. and Kant, V. (2018) 'Sociotechnical resilience: a preliminary concept', *Risk Analysis*, 38(1): 8–16.

Amoore, L. (2013) *The Politics of Possibility: Risk and Security beyond Probability*, Duke University Press.

Amoore, L. and Raley, R. (2017) 'Securing with algorithms: knowledge, decision, sovereignty', *Security Dialogue*, 48(1): 3–10.

Aradau, C. (2010) 'Security that matters: critical infrastructure and objects of protection', *Security Dialogue*, 41(5): 491–514.

Aradau, C. (2023) 'Algorithmic governmentality: questions of method', in W. Walters and M. Tazzioli (eds) *Handbook on Governmentality*, Edward Elgar, pp 235–50.

Aradau, C. and Van Munster, R. (2007) 'Governing terrorism through risk: taking precautions, (un)knowing the future', *European Journal of International Relations*, 13(1): 89–115.

Aradau, C. and Van Munster, R. (2008a) 'Insuring terrorism, assuring subjects, ensuring normality: the politics of risk after 9/11', *Alternatives: Global, Local, Political*, 33(2): 191–210.

Aradau, C. and Van Munster, R. (2008b) 'Taming the future: the dispositif of risk in the war on terror', in L. Amoore and M. De Goede (eds) *Risk and the War on Terror*, Routledge, pp 23–40.

Aradau, C. and Van Munster, R. (2011) *Politics of Catastrophe: Genealogies of the Unknown*, Routledge.

Ash, L. (2024) 'Thanks to a shadowy hacker group, the British Library is still on its knees: is there any way to stop them?', *The Guardian*, 6 February, Available from: https://www.theguardian.com/commentisfree/2024/feb/06/hacker-british-library-cybersecurity-cybercrime-uk

Ashenden, D. (2021) 'The future human and behavioural challenges of cybersecurity', in P. Cornish (ed) *The Oxford Handbook of Cybersecurity*, Oxford University Press, pp 723–34.

Avant, D., Finnemore, M. and Sell, S.K. (2010) 'Who governs the globe?', in D. Avant, M. Finnemore and S.K. Sell (eds) *Who Governs the Globe?*, Cambridge University Press, pp 1–32.

Aven, T. (2012) 'On the critique of Beck's view on risk and risk analysis', *Safety Science*, 50(4): 1043–8.

Aven, T. (2025) 'An uncertainty-based risk perspective on risk perception and communication: opportunities for new empirical-based research', *Risk Analysis*, 45(8): 2073–8.

Backman, S. (2021) 'Conceptualizing cyber crises', *Journal of Contingencies and Crisis Management*, 29(4): 429–38.

Backman, S. (2023) 'Risk vs. threat-based cybersecurity: the case of the EU', *European Security*, 32(1): 85–103.

Backman, S. and Stevens, T. (2024) 'Cyber risk logics and their implications for cybersecurity', *International Affairs*, 100(6): 2441–60.

Baker, T. and Simon, J. (2002) 'Embracing risk', in T. Baker and J. Simon (eds) *Embracing Risk: The Changing Culture of Insurance and Responsibility*, University of Chicago Press, pp 1–25.

Baker, T. and Shortland, A. (2023) 'Insurance and enterprise: cyber insurance for ransomware', *The Geneva Papers on Risk and Insurance: Issues and Practice*, 48: 275–99.

Balzacq, T. (2005) 'The three faces of securitization: political agency, audience and context', *European Journal of International Relations*, 11(2): 171–201.

Balzacq, T. (ed) (2010) *Securitisation Theory: How Security Problems Emerge and Dissolve*, Routledge.

Balzacq, T., Léonard, S. and Ruzicka, J. (2016) '"Securitization" revisited: theory and cases', *International Relations*, 30(4): 494–531.

Baram, G. and Peer, N. (2024) 'Cyberwarfare norms and the attribution imperative: shaping responsible state behaviour in cyberspace', in T. Stevens and J. Devanny (eds) *Research Handbook on Cyberwarfare*, Edward Elgar, pp 371–89.

Barnard-Wills, D. and Ashenden, D. (2012) 'Securing virtual space: cyber war, cyber terror, and risk', *Space and Culture*, 15(2): 110–23.

Barrinha, A. and Da Mota, S. (2017) 'Drones and the uninsurable security subjects', *Third World Quarterly*, 38(2): 253–69.

Barrinha, A. and Renard, T. (2020) 'Power and diplomacy in the post-liberal cyberspace', *International Affairs*, 96(3): 749–66.

Bartelson, J. (2009) 'Is there a global society?', *International Political Sociology*, 3(1): 112–15.

Basaran, T., Bigo, D., Guittet, E-P. and Walker, R.B.J. (2016) 'Transversal lines: an introduction', in T. Basaran, D. Bigo, E-P. Guittet and R.B.J. Walker (eds) *International Political Sociology: Transversal Lines*, Routledge, pp 1–9.

Battistelli, F. and Galantino, M.G. (2019) 'Dangers, risks and threats: an alternative conceptualization to the catch-all concept of risk', *Current Sociology*, 67(1): 64–78.

Beck, U. (1992) *Risk Society: Towards a New Modernity*, Sage.

Beck, U. (1998) *World Risk Society*, Polity.

Beck, U. (2009) *World at Risk*, Polity.

Behrent, M.C. (2013) 'Foucault and technology', *History and Technology*, 29(1): 54–104.

Bell, C. (2006) 'Surveillance strategies and populations at risk: biopolitical governance in Canada's national security policy', *Security Dialogue*, 37(2): 147–65.

Bellanova, R., Jacobsen K.L. and Monsees L. (2020) Taking the trouble: science, technology and security studies', *Critical Studies on Security*, 8(2): 87–100.

Benincasa, E. (2020) 'The role of regional organizations in building cyber resilience: ASEAN and the EU', *Pacific Forum Issues and Insights*, 20(3), Available from: https://pacforum.org/publications/issues-insights-vol-20-wp-3-the-role-of-regional-organizations-in-building-cyber-resilience-asean-and-the-eu/

Bernstein, P.L. (1996) *Against All Odds: The Remarkable Story of Risk*, John Wiley.

Betz, D.J. and Stevens, T. (2013) 'Analogical reasoning and cyber security', *Security Dialogue*, 44(2): 147–64.

Bialasiewicz, L., Campbell, D., Elden, S., Graham, S., Jeffrey, A. and Williams, A.J. (2007) 'Performing security: the imaginative geographies of current US strategy', *Political Geography*, 26(4): 405–22.

Bigo, D. (2002) 'Security and immigration: toward a critique of the governmentality of unease', *Alternatives*, 27(1, supplement): 63–92.

Bigo, D. (2008) 'Globalized (in)security: the field and the ban-opticon', in D. Bigo and A. Tsoukala (eds) *Terror, Insecurity and Liberty: Illiberal Practices of Liberal Regimes after 9/11*, Routledge, pp 10–48.

Bigo, D. and Walker, R.B.J. (2007) 'Political sociology and the problem of the international', *Millennium: Journal of International Studies*, 35(3): 725–39.

Björck, F., Henkel, M., Stirna, J. and Zdravkovic, J. (2015) 'Cyber resilience: fundamentals for a definition', in A. Rocha, A.M. Correia, S. Costanzo and L.P. Reis (eds) *New Contributions in Information Systems and Technologies*, vol 1, Springer, pp 311–16.

Blythe, J., Koppel R. and Smith, R.W. (2013) 'Circumvention of security: good users do bad things', *IEEE Security and Privacy*, 11(5): 80–3.

Boakye, D., Sarpong, D., Meissner, D. and Ofusu, G. (2024) 'How TalkTalk did the walk-walk: strategic reputational repair in a cyber-attack', *Information Technology and People*, 37(4): 1642–73.

Böhm, M. (2025) 'Algorithms, AI, big data, and big tech: IPS scholarship on digital technologies', *International Political Sociology*, 19(3): olaf019.

Bonfanti, M.E. (2022) 'Artificial intelligence and the offense–defense balance in cyber security', in M. Dunn Cavelty and A. Wenger (eds) *Cyber Security Politics: Socio-technological Transformations and Political Fragmentation*, Routledge, pp 64–79.

Borgmann, A. (1999) *Holding on to Reality: The Nature of Information at the Turn of the Millennium*, University of Chicago Press.

Bourbeau, P. (2018) *On Resilience: Genealogy, Logics, and World Politics*, Cambridge University Press.

Branch, J. (2020) 'What's in a name? Metaphors and cybersecurity', *International Organization*, 75(1): 39–70.

Branch, J. (2024) 'Territory, sovereignty and boundaries in digital battlespace', in T. Stevens and J. Devanny (eds) *Research Handbook on Cyberwarfare*, Edward Elgar, pp 301–15.

Brantly, A.F. (2016) *The Decision to Attack: Military and Intelligence Cyber Decision-Making*, University of Georgia Press.

Brantly, A.F. (2021) 'Risk and uncertainty can be analyzed in cyberspace', *Journal of Cybersecurity*, 7(1): tyab001.

British Library (2024) 'Learning lessons from the cyber-attack: British Library cyber incident review', Available from: https://www.bl.uk/home/british-library-cyber-incident-review-8-march-2024.pdf/

Broeders, D. (2021) 'Private active cyber defense and (international) cyber security: pushing the line?', *Journal of Cybersecurity*, 7(1): tyab010.

Broeders, D., Sukumar, A., Kello, M. and Andersen, L.H. (2025) 'Digital corporate autonomy: geo-economics and corporate agency in conflict and competition', *Review of International Political Economy*, 32(4): 1189–1213.

Brown, D. (2024) 'Incident response exercising: your prescription for building cyber resilience', NCC Group, 22 April, Available from: https://www.nccgroup.com/us/incident-response-exercises-for-building-cyber-resilience/

Buchanan, B. (2016) *The Cybersecurity Dilemma: Hacking, Trust, and Fear between Nations*, Hurst.

Buchanan, B. (2020) *The Hacker and the State: Cyber Attacks and the New Normal of Geopolitics*, Harvard University Press.

Bueger, C. (2016) 'Security as practice', in M. Dunn Cavelty and T. Balzacq (eds) *Routledge Handbook of Security Studies*, Routledge, pp 126–36.

Bueger, C., Liebetrau T. and Stockbruegger J. (2023) 'Theorizing infrastructures in global politics', *International Studies Quarterly*, 67(4): sqad101.

Burchell, G., Gordon, C. and Miller, P. (eds) (1991) *The Foucault Effect: Studies in Governmentality, with Two Lectures by and an Interview with Michel Foucault*, Chicago University Press.

Burgess, A. (2016) 'Introduction', in A. Burgess, A. Alemanno and J.O. Zinn (eds) *Routledge Handbook of Risk Studies*, Routledge, pp 1–14.

Burgess, J.P. (2007) 'Social values and material threat: the European Programme for Critical Infrastructure Protection', *International Journal of Critical Infrastructures*, 3(3–4): 471–87.

Burke, A. (2002) 'The aporias of security', *Alternatives*, 27(1): 1–27.

Burton, J. and Lain, C. (2020) 'Desecuritising cybersecurity: towards a societal approach', *Journal of Cyber Policy*, 5(3): 449–70.

Burton, J. and Christou, G. (2021) 'Bridging the gap between cyberwar and cyberpeace', *International Affairs*, 97(6): 1727–47.

Buzan, B., Wæver, O. and De Wilde, J. (1998) *Security: A New Framework for Analysis*, Lynne Rienner.

Buzatu, A.-M. (2022) 'Advanced Persistent Threat groups increasingly destabilize peace and security in cyberspace', in S.J. Shackleford, F. Douzet and C. Ankersen (eds) *Cyber Peace: Charting a Path toward a Sustainable, Stable, and Secure Cyberspace*, Cambridge University Press, pp 236–42.

Cains, M.G., Flora, L., Taber, D., King, Z. and Henshel, D.S. (2022) 'Defining cyber security and cyber security risk within a multidisciplinary context using expert elicitation', *Risk Analysis*, 42(8): 1643–69.

Camillo, M. (2017) 'Cyber risk and the changing role of insurance', *Journal of Cyber Policy*, 2(1): 53–63.

Campbell, E. (2019) 'Three-dimensional security: layers, spheres, volumes, milieus', *Political Geography*, 69: 10–21.

Canadian Centre for Cyber Security (2024) 'National cyber threat assessment 2025–2026', Available from: https://www.cyber.gc.ca/sites/default/files/national-cyber-threat-assessment-2025-2026-e.pdf

Carr, M. (2016) 'Public–private partnerships in national cyber-security strategies', *International Affairs*, 92(1): 43–62.

Carr, M. and Lesniewska, F. (2020) 'Internet of Things, cybersecurity and governing wicked problems: learning from climate change governance', *International Relations*, 34(3): 391–412.

Carrapico, H. and Farrand, B. (2024) 'Cybersecurity trends in the European Union: regulatory mercantilism and the digitalisation of geopolitics', *Journal of Common Market Studies*, 62: 147–58.

CASE Collective (2006) 'Critical approaches to security in Europe: a networked manifesto', *Security Dialogue*, 37(4): 443–87.

Chandler, D. (2013) 'Resilience and the autotelic subject: toward a critique of the societalization of security', *Security Dialogue*, 43(3): 213–29.

Chandler, D. (2018a) *Ontopolitics in the Anthropocene: An Introduction to Mapping, Sensing and Hacking*, Routledge.

Chandler, D. (2018b) 'Resilience: the societalization of security', in D. Chandler and J. Reid (eds) *The Neoliberal Subject: Resilience, Adaptation and Vulnerability*, Rowman and Littlefield, pp 27–50.

Christou, G. (2016) *Cybersecurity in the European Union: Resilience and Adaptability in Governance Policy*, Palgrave Macmillan.

Christou, G. (2019) 'The collective securitisation of cyberspace in the European Union', *West European Politics*, 42(2): 278–301.

Ciută, F. (2010) 'Conceptual notes on energy security: total or banal security?', *Security Dialogue*, 41(2): 123–44.

Clapton, W. (2014) *Risk and Hierarchy in International Society: Liberal Interventionism in the Post-Cold War Era*, Palgrave Macmillan.

Cobos, E.V., Ckir, S., Mei-Zahav, H. and Barakcin, B.B. (2024) 'The role of cybersecurity in economic performance', World Bank White Paper, Available from: https://documents1.worldbank.org/curated/en/099092324164526526/pdf/P178769189c7360111ac1f1185e04824dec.pdf

Cohen, J.E. (2007) 'Cyberspace as/and space', *Columbia Law Review*, 107(1): 210–56.

Coker, C. (2013) *War in an Age of Risk*, Polity.

Collier, J. (2018) 'Cyber security assemblages: a framework for understanding the dynamic and contested nature of security provision', *Politics and Governance*, 6(2): 13–21.

Collier, S.J. (2009) 'Topologies of power: Foucault's analysis of political government beyond "governmentality"', *Theory, Culture and Society*, 26(6): 78–108.

Corry, O. (2012) 'Securitisation and "riskification": second-order security and the politics of climate change', *Millennium: Journal of International Studies*, 40(2): 235–58.

Côté, A. (2016) 'Agents without agency: assessing the role of the audience in securitization theory', *Security Dialogue*, 47(6): 541–58.

Cox, R. (1981) 'Social forces, states and world orders: beyond international relations theory', *Social Forces*, 10(2): 126–55.

Creese, S. and Joshi, A. (2024) 'Unpacking cyber resilience', University of Oxford and World Economic Forum White Paper, Available from: https://gcscc.ox.ac.uk/sitefiles/wef-unpacking-cyber-resilience-2024.pdf

Cremer, F., Sheehan, B., Mullins, M., Fortmann, N., Materne S. and Murphy F. (2024a) 'Enhancing cyber insurance strategies: exploring reinsurance and alternative risk transfer approaches', *Journal of Cybersecurity*, 10(1): tyae027.

Cremer, F., Sheehan, B., Mullins, M., Fortmann, N., Ryan B.J. and Materne S. (2024b) 'On the insurability of cyber warfare: an investigation into the German cyber insurance market', *Computers and Security*, 142: 103886.

Cristiano, F., Kurowska, X., Stevens, T., Hurel, L.M., Fouad, N.S., Dunn Cavelty, M. et al (2023) 'Cybersecurity and the politics of knowledge production: toward a reflexive practice', *Journal of Cyber Policy*, 8(3): 331–64.

Crowcroft, J. (2010) 'Internet failures: an emergent sea of complex systems and critical design errors?', *The Computer Journal*, 53(1): 1752–7.

Cunningham, K. and Tomes, R.R. (2004) 'Space-time orientations and contemporary political-military thought', *Armed Forces and Society*, 31(1): 119–40.

Dalby, S. (2013) 'Biopolitics and climate security in the Anthropocene', *Geoforum*, 49: 184–92.

Da Silva, J. (2022) 'Teacher, enforcer, soothsayer, scapegoat: the purpose of the CISO in commercial organisations', PhD thesis, Royal Holloway University of London, Available from: https://pure.royalholloway.ac.uk/en/publications/teacher-enforcer-soothsayer-scapegoat-the-purpose-of-the-ciso-in-/

Da Silva, J. and Jensen, R.B. (2022) '"Cyber security is a dark art": the CISO as soothsayer', *Proceedings of the ACM on Human-Computer Interaction* 6 (CSCW2), Available from: https://doi.org/10.1145/3555090

Dean, M.M. (1998) 'Risk, calculable and incalculable', *Soziale Welt*, 49(1): 25–42.

Dean, M.M. (2010) *Governmentality: Power and Rule in Modern Society* (2nd edn), Sage.

Deibert, R.J. and Rohozinski, R. (2010) 'Risking security: policies and paradoxes of cyberspace security', *International Political Sociology*, 4(1): 15–32.

DeNardis, L. (2009) *Protocol Politics: The Globalization of Internet Governance*, MIT Press.

Department for Science, Innovation and Technology (DSIT) (2025) 'Cyber security breaches survey 2025', Available from: https://www.gov.uk/government/statistics/cyber-security-breaches-survey-2025/cyber-security-breaches-survey-2025

Deuchars R. (2004) *The International Political Economy of Risk: Rationalism, Calculation and Power*, Ashgate.

Devanny, J. (2022) '"Madman theory" or "persistent engagement"? The coherence of US cyber strategy under Trump', *Journal of Applied Security Research*, 17(3): 282–309.

Dhillon, G. and Backhouse, J. (2001) 'Current directions in IS security research: towards socio-organizational perspectives', *Information Systems Journal*, 11(2): 127–53.

Dillon, M.C. (2007) 'Governing terror: the state of emergency of biopolitical emergence', *International Political Sociology*, 1(1): 7–28.

Dillon, M.C. (2008) 'Underwriting security', *Security Dialogue*, 39(2–3): 309–32.

Dillon, M.C. (2015) *Biopolitics of Security: A Political Analytic of Finitude*, Routledge.

Dillon, M.C. and Lobo-Guerrero, L. (2008) 'Biopolitics of security in the 21st century: an introduction', *Review of International Studies*, 34(2): 265–92.

Dimitriadis, C. (2025) 'From the cyber trenches: expert recommendations for resilience policy', National Preparedness Commission, 10 January, Available from: https://nationalpreparednesscommission.uk/publications/from-the-cyber-trenches-expert-recommendations-for-resilience-policy/

Dittmer, J. (2017) *Diplomatic Material: Affect, Assemblage, and Foreign Policy*, Duke University Press.

Dossi, S. (2020) 'On the asymmetric advantages of cyberwarfare: Western literature and the Chinese journal *Guofang Keji*', *Journal of Strategic Studies*, 43(2): 281–308.

Duffield, M. (2007) *Development, Security and Unending War: Governing the World of Peoples*, Polity.

Dunn Cavelty, M. (2007) *Cyber-Security and Threat Politics: US Efforts to Secure the Information Age*, Routledge.

Dunn Cavelty, M. (2008) 'Like a phoenix from the ashes: the reinvention of infrastructure protection as distributed security', in M. Dunn Cavelty and K.S. Kristensen (eds) *Securing 'the Homeland': Critical Infrastructure, Risk and (In)Security*, Routledge, pp 40–62.

Dunn Cavelty, M. (2013) 'From cyber-bombs to political fallout: threat representations with an impact in the cyber-security discourse', *International Studies Review*, 15(1): 105–22.

Dunn Cavelty, M. (2017) 'Cyber-security and private actors', in A. Leander and R. Abrahamsen (eds) *Routledge Handbook of Private Security Studies*, Routledge, pp 89–99.

Dunn Cavelty, M. (2020) 'Cybersecurity between hypersecuritization and technological routine', in E. Tikk and M. Kerttunen (eds) *Routledge Handbook of International Cybersecurity*, Routledge, pp 11–21.

Dunn Cavelty, M. (2025) 'Digital capabilities as a strategic resource', in G. Grgić and D. Möckli (eds) *Strategic Trends 2025: Key Developments in Global Affairs*, ETH Zurich, pp 95–115.

Dunn Cavelty, M. and Egloff, F.J. (2019) 'The politics of cybersecurity', *St. Anthony's International Review*, 15(1): 37–57.

Dunn Cavelty, M., Kaufmann, M. and Kristensen, K.S. (2015) 'Resilience and (in)security: practices, subjects, temporalities', *Security Dialogue*, 46(1): 3–14.

Dunn Cavelty, M., Eriksen, C. and Scharte, B. (2023) 'Making cyber security more resilient: adding social considerations to technological fixes', *Journal of Risk Research*, 26(7): 801–14.

Dupont, B. (2019) 'The cyber-resilience of financial institutions: significance and applicability', *Journal of Cybersecurity*, 5(1): tyz013.

Dwyer, A.C. (2023) 'Cybersecurity's grammars: a more-than-human geopolitics of computation', *Area*, 55(1): 10–17.

Dwyer, A.C., Stevens, C., Muller, L.P., Dunn Cavelty, M., Coles-Kemp, L. and Thornton, P. (2022) 'What can a critical cybersecurity do?', *International Political Sociology*, 16(3), olac013.

Ebert, N., Schaltegger, T., Ambuehl, B., Schöni, L., Zimmermann V. and Knieps, M. (2023) 'Learning from safety science: a way forward for studying cybersecurity incidents in organizations', *Computers and Security*, 134: 103435.

Elden, S. (2013) 'Secure the volume: vertical geopolitics and the depth of power', *Political Geography*, 34: 35–51.

Eling, M. and Schnell, W. (2016) 'What do we know about cyber risk and cyber risk insurance?', *Journal of Risk Finance*, 17(5): 474–91.

Esmark, A. (2020) *The New Technocracy*, Bristol University Press.

European Union Agency for Cybersecurity (2025) 'Space threat landscape 2025', Available from: https://www.enisa.europa.eu/publications/enisa-space-threat-landscape-2025

Evans, A. (2019) *Managing Cyber Risk*, Routledge.

Evans, B. and Reid, J. (2013) 'Dangerously exposed: the life and death of the resilient subject', *Resilience: International Policies, Practices and Discourses*, 1(2): 83–98.

Evans, B. and Reid, J. (2014) *Resilient Life: The Art of Living Dangerously*, Polity.

Evans, B. and Reid, J. (2015) 'Exhausted by resilience: response to the commentaries', *Resilience: International Policies, Practices and Discourses*, 3(2): 154–9.

Evans, S.W., Leese, M. and Rychnovská, D. (2021) 'Science, technology, security: towards critical collaboration', *Social Studies of Science*, 51(2): 189–213.

Ewald, F. (1991) 'Insurance and risk', in G. Burchell, C. Gordon and P. Miller (eds) *The Foucault Effect: Studies in Governmentality, with Two Lectures by and an Interview with Michel Foucault*, Chicago University Press, pp 197–210.

Falco, G. and Rosenbach, E. (2022) *Confronting Cyber Risk: An Embedded Endurance Strategy for Cybersecurity*, Oxford University Press.

Farrand, B., Carrapico, H. and Turobov, A. (2024) 'The new geopolitics of EU cybersecurity: security, economy and sovereignty', *International Affairs*, 100(6): 2379–97.

Farrand, B., Carrapico, H., Massoumi, N., McGowan, W. and Mythen, G. (2020) 'Disputing security and risk: the convoluted politics of uncertainty', in I. Scoones and A. Stirling (eds) *The Politics of Uncertainty: Challenges of Transformation*, Routledge, pp 151–63.

Feyerabend, P. (1975) *Against Method: Outline of an Anarchistic Theory of Knowledge*, Verso Books.

Fischerkeller, M.P., Goldman, E.O. and Harknett, R.J. (2023) *Cyber Persistence Theory: Redefining National Security in Cyberspace*, Oxford University Press.

Forscey, D., Bateman, J., Beecroft, N. and Woods, B. (2022) 'Systemic cyber risk: a primer', Carnegie Endowment for international Peace, Available from: https://carnegieendowment.org/2022/03/07/systemic-cyber-risk-primer-pub-86531

Fouad, N.S. (2022) 'The non-anthropocentric informational agents: codes, software, and the logic of emergence in cybersecurity', *Review of International Studies*, 48(4): 766–85.

Fouad, N.S. (2025a) *Theorising Cyber In(Security): Information, Materiality, and Entropic Security*, Routledge.

Fouad, N.S. (2025b) 'Failing better in the infosphere: ontological uncertainties and the essence of security in cyberspace', *Information, Communication and Society*, Available from: https://doi.org/10.1080/1369118X.2025.2492574

Foucault, M. (1977) *Discipline and Punish: The Birth of the Prison*, Allen Lane.

Foucault, M. (1991) 'Governmentality', in G. Burchell, C. Gordon and P. Miller (eds) *The Foucault Effect: Studies in Governmentality, with Two Lectures by and an Interview with Michel Foucault*, Chicago University Press, pp 87–104.

Foucault, M. (1997) 'What is critique?', in S. Lotringer (ed) *The Politics of Truth: Michel Foucault*, Semiotext(e), pp 41–81.

Friedman, J. (2025) 'The world is more uncertain than you think: assessing and combating overconfidence among 2000 national security officials', *Texas National Security Review*, 8(4): 34–48.

Friis, K. and Reichborn-Kjennerud, E. (2016) 'From cyber-threats to cyber risks', in K. Friis and J. Ringsmose (eds) *Conflict in Cyberspace: Theoretical, Strategic and Legal Perspectives*, Routledge, pp 27–44.

Gallagher Re (2024) 'Protecting the digital revolution: the state of the Asian cyber insurance market in 2024', Available from: https://www.ajg.com/gallagherre/-/media/files/gallagher/gallagherre/news-and-insights/2024/september/gallagherre-protecting-the-digital-revolution.pdf

Gallie, W.B. (1955) 'Essentially contested concepts', *Proceedings of the Aristotelian Society*, 56: 167–98.

Galloway, A.R. and Thacker, E. (2007) *Exploit: A Theory of Networks*, University of Minnesota Press.

Giddens, A. (1990) *The Consequences of Modernity*, Polity.

Giddens, A. (1991) *Modernity and Self-identity: Self and Society in the Late Modern Age*, Polity.

Giles, K. (2021) 'Russian information warfare: construct and purpose', in T. Clack and R. Johnson (eds) *The World Information War: Western Resilience, Campaigning, and Cognitive Effects*, Routledge, pp 139–61.

Gioe, D.V., Goodman, M.S. and Stevens, T. (2020) 'Intelligence in the cyber era: evolution or revolution?', *Political Science Quarterly*, 135(2): 191–224.

Global Federation of Insurance Associations (2024) 'Annual report 2023–24', Available from: https://reports.gfiainsurance.org/gfia-annual-report-2023-2024/

Goldin, I. and Mariathasan, M. (2014) *The Butterfly Defect: How Globalization Creates Systemic Risks, and What to Do about It*, Princeton University Press.

Gomez, M.A. and Villar, E.B. (2018) 'Fear, uncertainty, and dread: cognitive heuristics and cyber threats', *Politics and Governance*, 6(2): 61–72.

Gould-Davies, N. (2019) *Tectonic Politics: Global Political Risk in an Age of Transformation*, Brookings Institution.

Graham, J.D. and Wiener, J.B. (2009) 'Confronting risk tradeoffs', in J.D. Graham and J.B. Wiener (eds) *Risk versus Risk: Tradeoffs in Protecting Health and the Environment*, Harvard University Press, pp 1–41.

Grove, K. (2018) *Resilience*, Routledge.

Hacking, I. (1990) *The Taming of Chance*, Cambridge University Press.

Haggart, B., Tusikov, N. and Scholte, J.A. (eds) (2021) *Power and Authority in Internet Governance: Return of the State?*, Routledge.

Hagmann, J. and Dunn Cavelty, M. (2012) 'National risk registers: security scientism and the propagation of permanent insecurity', *Security Dialogue*, 43(1): 79–96.

Hall, P., Heath, C. and Coles-Kemp, L. (2015) 'Critical visualization: a case for rethinking how we visualize risk and security', *Journal of Cybersecurity*, 1(1): 93–108.

Hansen, L. and Nissenbaum, H. (2009) 'Digital disaster, cyber security, and the Copenhagen School', *International Studies Quarterly*, 53(4): 1155–75.

Hansen, S.F., Von Krauss, M.K and Tickner, J.A. (2008) 'The precautionary principle and risk–risk tradeoffs', *Journal of Risk Research*, 11(4): 423–64.

Hardy, A. and Le Blanc, N. (2023) 'Securing tomorrow: why Latin America should top global cyber insurers' lists', DAC Beachcroft, 6 December, Available from: https://www.dacbeachcroft.com/en/What-we-think/Securing-Tomorrow-Why-Latin-America-should-top-Global-Cyber-Insurers-Lists

Harknett, R.J. (2020) 'SolarWinds: the need for persistent engagement', Lawfare, 23 December, Available from: https://www.lawfaremedia.org/article/solarwinds-need-persistent-engagement

Harknett, R.J. and Smeets, M. (2022) 'Cyber campaigns and strategic outcomes', *Journal of Strategic Studies*, 45(4): 534–67.

Harknett, R.J. and Stever, J.A. (2009) 'The cybersecurity triad: government, private sector partners, and the engaged cybersecurity citizen', *Journal of Homeland Security and Emergency Management*, 6(1): article 79.

Hassib, B. and Shires, J. (2021) 'Manipulating uncertainty: cybersecurity politics in Egypt', *Journal of Cybersecurity*, 7(1): tyaa026.

Healey, J. (2011) 'Pursuing cyber statecraft', Atlantic Council Issue Brief, 13 September, Available from: https://www.atlanticcouncil.org/in-depth-research-reports/issue-brief/pursuing-cyber-statecraft/

Heath-Kelly, C. (2015) 'Securing through the failure to secure? The ambiguity of resilience at the bombsite', *Security Dialogue*, 46(1): 69–85.

Heidrick and Struggles (2023) 'Global Chief Information Security Officer (CISO) survey', June, Available from: https://www.heidrick.com/-/media/heidrickcom/publications-and-reports/2023-global-chief-information-security-officer-survey.pdf

Heng, Y-K. (2006) *War as Risk Management: Strategy and Conflict in an Age of Globalised Risks*, Routledge.

Heng, Y-K. (2018) 'The continuing resonance of the war as risk management perspective for understanding military interventions', *Contemporary Security Policy*, 39(4): 544–58.

Herr, T. (2021) 'Cyber insurance and private governance: the enforcement power of markets', *Regulation and Governance*, 15(1): 98–114.

HM Government (2021) 'National cyber strategy 2022: pioneering a cyber future with the whole of the UK', Available from: https://assets.publishing.service.gov.uk/government/uploads/system/uploads/attachment_data/file/1053023/national-cyber-strategy-amend.pdf

HM Government (2022) 'Government cyber security strategy 2022 to 2030: building a cyber resilient public sector', Available from: https://www.gov.uk/government/publications/government-cyber-security-strategy-2022-to-2030

HM Government (2025a) 'Emergency preparedness, resilience and response concept of operations', 15 January, Available from: https://www.gov.uk/government/publications/emergency-preparedness-resilience-and-response-concept-of-operations/emergency-preparedness-resilience-and-response-concept-of-operations

HM Government (2025b) 'Cyber security and resilience policy statement', 9 April, Available from: https://www.gov.uk/government/publications/cyber-security-and-resilience-bill-policy-statement/cyber-security-and-resilience-bill-policy-statement

HM Government (2025c) 'The Strategic Defence Review 2025: Making Britain safer; secure at home, strong abroad', 11 June, Available from: https://www.gov.uk/government/publications/the-strategic-defence-review-2025-making-britain-safer-secure-at-home-strong-abroad

Hill, S. and Creese, S. (2021) 'Cyber resilience: a leadership challenge and responsibility', National Preparedness Commission, 14 November, Available from: https://nationalpreparednesscommission.uk/publications/cyber-resilience-a-leadership-challenge-and-responsibility/

Hoffmann, A. (2021) 'The transnational and the international: from critique of statism to transversal lines', *Cambridge Review of International Affairs*, 35(6): 796–810.

Holmström, A. and Große, C. (2025) 'Not all heroes wear capes: cyber resilience of the social administration at a Swedish municipality', *Risk, Hazards, and Crisis in Public Policy*, 16(3): e70024.

Hom, A.R. (2020) *International Relations and the Problem of Time*, Oxford University Press.

Hörnqvist, M. (2010) *Risk, Power and the State: After Foucault*, Routledge.

Huang, K., Madnick, S., Choucri, N. and Zhang F. (2021) 'A systematic framework to understand transnational governance for cybersecurity risks from digital trade', *Global Policy*, 12(5): 625–38.

Hubbard, G.A. (2023) 'State-level cyber resilience: a conceptual framework', *Applied Cybersecurity and Internet Governance*, 2(1), Available from: https://doi.org/10.60097/ACIG/162859

Hurel, L.M. (2025) 'The challenge of non-Western cyber attribution', RUSI video commentary, 29 May, Available from: https://www.rusi.org/news-and-comment/video-commentary/challenge-non-western-cyber-attribution

Huysmans, J. (2004) 'Minding exceptions: the politics of insecurity and liberal democracy', *Contemporary Political Theory*, 3(3): 321–41.

Huysmans, J. (2011) 'What's in an act? On security speech acts and little security nothings', *Security Dialogue*, 42(4–5): 371–83.

Huysmans, J. and Nogueira, J.P. (2016) 'Ten years of IPS: fracturing IR', *International Political Sociology*, 10(4): 299–319.

Huysmans, J. and Nogueira, J.P. (2021) 'International political sociology as a mode of critique: fracturing totalities', *International Political Sociology*, 15(1): 2–21.

Iasiello, E. (2024) 'Blind allegations of US cyber hegemony complements China's influence campaign to weaken global confidence in the United States', *The Cyber Defense Review*, 9(3): 37–49.

Institute of Risk Management (nd) 'Cyber risk', Available from: https://www.theirm.org/what-we-say/thought-leadership/cyber-risk/

Jabbour, K. and Poisson, J. (2016) 'Cyber risk assessment in distributed information systems', *The Cyber Defense Review*, 1(1): 91–112.

Jackson, R. (2015) 'The epistemological crisis of counterterrorism', *Critical Studies on Terrorism*, 8(1): 33–54.

Jalabert, Q., Van Puyvelde, D. and Maguire, T. (2025) 'Calling out Russia: France's shift to public attribution', War on the Rocks, 3 July, Available from: https://warontherocks.com/2025/07/calling-out-russia-frances-shift-on-public-attribution/

Jarjoui, S., Murimi, R. and Murimi, R. (2024) 'Communities, agency, and resilience: a perspective addressing tragedy of the cyber commons', *The Cyber Defense Review*, 9(1): 113–31.

Jarvis, L. (2019) 'Toward a vernacular security studies: origins, interlocutors, contributions, and challenges', *International Studies Review*, 21(1): 107–26.

Jensen, B.M., Whyte, C. and Cuomo, S. (2020) 'Algorithms at war: the promise, peril, and limits of artificial intelligence', *International Studies Review*, 22(3): 526–50.

Johansmeyer, T. (2023) 'If cyber is uninsurable, the United States has a major strategy problem', Lawfare, 26 July, Available from: https://www.lawfaremedia.org/article/if-cyber-is-uninsurable-the-united-states-has-a-major-strategy-problem

Joinson, A., Dixon, M., Coventry, L. and Briggs, P. (2023) 'Development of a new "human cyber-resilience scale"', *Journal of Cybersecurity*, 9(1): tyad007.

Jones, C. (2025) 'Supply chain attacks surge with orgs "flying blind" about dependencies', The Register, 25 June, Available from: https://www.theregister.com/2025/06/25/supply_chain_attacks_hammer_organizations

Joseph, J. (2018) *Varieties of Resilience: Studies in Governmentality*, Cambridge University Press.

Joseph, J. (2021) 'Variations of governmentality across the globe: the case of resilience', in J. Busse (ed) *The Globality of Governmentality: Governing an Entangled World*, Routledge, pp 50–67.

Kaminska, M. (2021) 'Restraint under conditions of uncertainty: why the United States tolerates cyberattacks', *Journal of Cybersecurity*, 7(1): tyab008.

Katzenstein, P.J. and Seybert, L.A. (2018) 'Protean power and control power: conceptual analysis', in P.J. Katzenstein and L.A. Seybert (eds) *Protean Power: Exploring the Uncertain and Unexpected in World Politics*, Cambridge University Press, pp 3–26.

Kaufmann, M. (2013) 'Emergent self-organisation in emergencies: resilience rationales in interconnected societies', *Resilience: International Policies, Practices and Discourses*, 1(1): 53–68.

Kaufmann, M. (2017) *Resilience, Emergencies and the Internet: Security In-Formation*, Routledge.

Khalili, M.M., Liu, M. and Romanosky, S. (2019) 'Embracing and controlling risk dependency in cyber-insurance policy underwriting', *Journal of Cybersecurity*, 5(1): tyz010.

Khando, K., Gao, S., Islam, S.M. and Salman, A. (2021) 'Enhancing employees information security awareness in private and public organisations: a systematic literature review', *Computers and Security*, 106: 102267.

Klepper, D. (2025) 'Iranian-backed hackers go to work after US strikes', ABC News, 25 June, Available from: https://apnews.com/article/iran-trump-cybersecurity-hacking-9009bff8425d97366e9423b50fb52edf

Klimburg, A. (2011) 'Mobilising cyber power', *Survival*, 53(1): 41–60.

Knight, F.H. (2009) [1921] *Risk, Uncertainty, and Profit*, Signalman Publishing.

Kostyuk, N. and Wayne, C. (2021) 'The microfoundations of state cybersecurity: cyber risk perceptions and the mass public', *Journal of Global Security Studies*, 6(2): ogz077.

Kostyuk, N. and Brantly, A.F. (2022) 'War in the borderland through cyberspace: limits of defending Ukraine through interstate cooperation', *Contemporary Security Policy*, 43(3): 498–515.

Kostyuk, N., Perkoski, E. and Poznansky, M. (2025) 'The bureaucratic politics of cyber strategy', *Journal of Global Security Studies*, 10(3): ogaf007.

Kott, A. and Linkov, I. (2021) 'To improve cyber resilience, measure it', *Computer*, 54(2): 80–5.

Kratochwil, F. (2011) 'Making sense of "international practices"', in E. Adler and V. Pouliot (eds) *International Practices*, Cambridge University Press, pp 36–60.

Kruck, A. and Weiss, M. (2025) 'Disentangling *Leviathan* on its home turf: authority foundations, policy instruments, and the making of security', *Regulation and Governance*, 19(1): 146–60.

Lakshmi, R., Naseer, H., Maynard, S. and Ahmad, A. (2021) 'Sensemaking in cybersecurity incident response: the interplay of organizations, technology and individuals', 29th European Conference on Information Systems (ECIS 2021), Available from: https://arxiv.org/abs/2107.02941

Lambach, D. (2020) 'The territorialization of cyberspace', *International Studies Review*, 22(3): 482–506.

Larner, W. and Walters, W. (2004) 'Globalization as governmentality', *Alternatives*, 29(5): 495–514.

Lawson, S.T. (2021) *Cybersecurity Discourse in the United States: Cyber-Doom Rhetoric and Beyond*, Routledge.

Lechner, S. and Frost, M. (2018) *Practice Theory and International Relations*, Cambridge University Press.

Lee, F. (2023) 'Cyber insurance rate hikes slow – but exclusions expand', *FT Professional*, 22 May, Available from: https://professional.ft.com/en-gb/blog/cyber-insurance-rate-hikes-slow-but-exclusions-expand/

Lemke, T. (2011) '"The birth of bio-politics": Michel Foucault's lecture at the Collège de France on neo-liberal governmentality', *Economy and Society*, 30(2): 190–207.

Lemnitzer, J.M. (2021) 'Why cybersecurity insurance should be regulated and compulsory', *Journal of Cyber Policy*, 6(2): 118–36.

Libicki, M.C. (2012) 'The specter of non-obvious warfare', *Strategic Studies Quarterly*, 6(3): 88–101.

Liebetrau, T. and Christensen, K.K. (2021) 'The ontological politics of cyber security: emerging agencies, actors, sites, and spaces', *European Journal of International Security*, 6(1): 25–43.

Lillen, S. (2025) 'Viewpoint: loss control services are underused tools to reduce identity security risk', *Insurance Journal*, 22 May, Available from: https://www.insurancejournal.com/news/national/2025/05/22/824706.htm

Lobo-Guerrero, L. (2010) 'The international political sociology of risk', in R.A. Denemark and R. Marlin-Bennett (eds) *The International Studies Encyclopedia*, Wiley-Blackwell, pp 4418–36.

Lobo-Guerrero, L. (2011) *Insuring Security: Biopolitics, Security and Risk*, Routledge.

Lobo-Guerrero, L. (2014) 'Life securitisation, the event object of insurance and the strategisation of time', *Journal of Cultural Economy*, 7(3): 353–70.

Lockton Re (2025) 'Cyber risk pools and public private partnerships: time to dive in?', Available from: https://global.lockton.com/re/en/news-insights/lockton-re-cyber-report-cyber-risk-pools-and-public-private-partnerships

Luhmann, N. (2005) [1991] *Risk: A Sociological Theory*, Aldine Transaction.

Lupton, D. (2013) *Risk* (2nd edn), Routledge.

Lyotard, J.-F. (1984) [1979] *The Postmodern Condition: A Report on Knowledge*, University of Minnesota Press.

MacColl, J., Nurse, J.R.C. and Sullivan, J. (2021) 'Cyber insurance and the cyber security challenge', RUSI Occasional Paper, Available from: https://rusi.org/explore-our-research/publications/occasional-papers/cyber-insurance-and-cyber-security-challenge

MacColl, J., Sullivan, J., Nurse, J.R.C., Mott, G., Turner, S., Cartwright, E. et al (2023) 'Cyber Insurance and the ransomware challenge', RUSI Occasional Paper, Available from: https://www.rusi.org/explore-our-research/publications/occasional-papers/cyber-insurance-and-ransomware-challenge

Maness, R.C., Valeriano, B., Hedgecock, K., Macias, J.M. and Jensen, B. (2023) 'Expanding the Dyadic Cyber Incident and Campaign Dataset (DCID): cyber conflict from 2000 to 2020', *The Cyber Defense Review*, 8(2): 65–90.

Martin, A. (2024a) 'British Library hailed by UK cyber agency for its response to ransomware attack', The Record, 18 March, Available from: https://therecord.media/british-library-incident-response-uk-ncsc-praise

Martin, A. (2024b) 'UK election campaigns ignore London hospitals cyberattack despite disruption', The Record, 7 June, Available from: https://therecord.media/uk-election-campaigns-ignore-london-attacks

Martin, A., Tidy, J., Zetter, K., Devanny, J. and Stevens, T. (2024) 'Reporting on cyberwarfare: a conversation', in T. Stevens and J. Devanny (eds) *Research Handbook on Cyberwarfare*, Edward Elgar, pp 67–78.

Martin, C. (2023) 'The development of the United Kingdom's cyber posture', in R. Chesney and M. Smeets (eds) *Deter, Disrupt, or Deceive: Assessing Cyber Conflict as an Intelligence Contest*, Georgetown University Press, pp 201–21.

Martin, C. (2025) 'Jaguar Land Rover to M&S: our cyber policy blind spot', Voices, Blavatnik School of Government, 15 September, Available from: https://www.bsg.ox.ac.uk/blog/jaguar-land-rover-ms-our-cyber-policy-blind-spot

Martin, P. and Giddings, J. (2020) 'Building better resilience', National Preparedness Commission, Available from: https://nationalpreparednesscommission.uk/publications/building-better-resilience/

Maschmeyer, L. (2024) *Subversion: From Covert Operations to Cyber Conflict*, Oxford University Press.

Massey, D. (2005) *For Space*, Sage.

Mathew, A. (2024) 'Unscripted practices for uncertain events: organizational problems in cybersecurity incident management', *Science, Technology, and Human Values*, 49(4): 827–50.

Mathur, R. (2025) 'Techno-capitalism and weaponization of cyberspace', *Global Studies Quarterly*, 5(2): ksaf031.

McCourt, D.M. (2016) 'Practice theory and relationalism as the new constructivism', *International Studies Quarterly*, 60(3): 475–85.

McLean, C., Patterson, A. and Williams, J. (2009) 'Risk assessment, policy-making and the limits of knowledge: the precautionary principle and international relations', *International Relations*, 23(4): 548–66.

Meland, P.H., Tøndel, I.A., Moe, M. and Seehusen, F. (2017) 'Facing uncertainty in cyber insurance policies', in G. Livraga and C. Mitchell (eds) *Security and Trust Management*, Springer, pp 89–100.

Moore, D. (2022) *Offensive Cyber Operations: Understanding Intangible Warfare*, Hurst.

Mott, G., Turner, S., Nurse, J.R.C., MacColl, J., Sullivan, J., Cartwright, A. et al (2023) 'Between a rock and a hard(ening) place: cyber insurance in the ransomware era', *Computers and Security*, 128: 103162.

Mottram, R. (2008) 'Protecting the citizen in the twenty-first century: issues and challenges', in P. Hennessy (ed) *The New Protective State*, Continuum, pp 42–65.

Moynihan, H. (2021) 'The vital roles of international law in the framework for responsible state behaviour in cyberspace', *Journal of Cyber Policy*, 6(3): 394–410.

Muller, B. (2010) *Security, Risk and the Biometric State: Governing Borders and Bodies*, Routledge.

Munich Re (2025) 'Cyber insurance: risks and trends 2025', 4 March, Available from: https://www.munichre.com/en/insights/cyber/cyber-insurance-risks-and-trends-2025.html

Munk, T. (2022) *The Rise of Politically Motivated Cyber Attacks: Actors, Attacks and Cybersecurity*, Routledge.

Nakasone, P.M. (2019) 'A cyber force for persistent operations', *Joint Force Quarterly*, 92(1): 10–14.

National Audit Office (2025) 'Government cyber resilience', Available from: https://www.nao.org.uk/wp-content/uploads/2025/01/government-cyber-resilience.pdf

National Cyber Force (2023) 'Responsible cyber power in practice', Available from: https://www.gchq.gov.uk/news/ncf-responsible-cyber-power-in-practice

National Cyber Resilience Centre Group (nd) 'Welcome to National CRC Group', Available from: https://nationalcrcgroup.co.uk/

National Cyber Security Centre (NCSC) (nd, a) 'Glossary', Available from: https://www.ncsc.gov.uk/section/advice-guidance/glossary

National Cyber Security Centre (NCSC) (nd, b) 'Exercise in a box', Available from: https://www.ncsc.gov.uk/section/exercise-in-a-box/tabletop-exercises

National Cyber Security Centre (NCSC) (2023a) 'Cyber Security Toolkit for boards', 30 March, Available from: https://www.ncsc.gov.uk/collection/board-toolkit

National Cyber Security Centre (NCSC) (2023b) 'Risk management', 23 June, Available from: https://www.ncsc.gov.uk/collection/risk-management

National Cyber Security Centre (NCSC) (2023c) 'Introducing system and component driven risk management approaches', 23 June, Available from: https://www.ncsc.gov.uk/collection/risk-management/introducing-system-and-component-driven-risk-management-approaches

National Cyber Security Centre (NCSC) (2023d) 'Categorising UK cyber incidents', 23 August, Available from: https://www.ncsc.gov.uk/information/categorising-uk-cyber-incidents

National Cyber Security Centre (NCSC) (2024a) 'Cyber Assessment Framework', 18 April, Available from: https://www.ncsc.gov.uk/collection/cyber-assessment-framework

National Cyber Security Centre (NCSC) (2024b) 'NCSC CEO's speech to mark the launch of the NCSC Annual Review 2024', 3 December, Available from: https://www.ncsc.gov.uk/speech/ncsc-annual-review-launch-2024-ceo-dr-richard-horne

Neal, A.W. (2019) *Security as Politics: Beyond the State of Exception*, Edinburgh University Press.

Nissenbaum, H. (2005) 'Where computer security meets national security', *Ethics and Information Technology*, 7(2): 61–73.

Nyman, J. (2021) 'The everyday life of security: capturing space, practice, and affect', *International Political Sociology*, 15(3): 313–37.

Obendiek, A.S. (2025) 'Private companies in digital security: how private control over information and knowledge shapes regulatory challenges', *Journal of Global Security Studies*, 10(3): ogaf016.

O'Grady, N. (2016) 'A politics of redeployment: malleable technologies and the localisation of anticipatory calculation', in L. Amoore and V. Piotukh (eds) *Algorithmic Life: Calculative Devices in the Age of Big Data*, Routledge, pp 72–86.

O'Malley, P. (2004) *Risk, Uncertainty and Government*, Glasshouse Press.

O'Malley, P. (2016) 'Governmentality and the analysis of risk', in A. Burgess, A. Alemanno and J. Zinn (eds) *Routledge Handbook of Risk Studies*, Routledge, pp 109–16.

Omand, D. (2010) *Securing the State*, Hurst.

Oosthoek, K. and Doerr, C. (2021) 'Cyber threat intelligence: a product without a process?', *International Journal of Intelligence and Counterintelligence*, 34(2): 300–15.

Oppenheimer, H. (2024) 'How the process of discovering cyberattacks biases our understanding of cybersecurity', *Journal of Peace Research*, 61(1): 28–43.

Organisation for Economic Co-operation and Development (OECD) (2017) 'Enhancing the role of insurance in cyber risk management', Available from: https://www.oecd.org/en/publications/enhancing-the-role-of-insurance-in-cyber-risk-management_9789264282148-en.html

Osborn, E. and Simpson, A. (2018) 'Risk and the small-scale cyber security decision making dialogue: a UK case study', *The Computer Journal*, 61(4): 472–95.

Pain, D. (2023) 'Cyber risk accumulation: fully tackling the insurability challenge', The Geneva Association, Available from: https://www.genevaassociation.org/sites/default/files/2023-11/cyber_accumulation_report_91123.pdf

Pawlak, P. and Delerue, F. (eds) (2022) *A Language of Power? Cyber Defence in the European Union*, Chaillot Paper 176, EU Institute for Security Studies, Available from: https://www.iss.europa.eu/publications/chaillot-papers/language-power

Perrow, C. (1999) *Normal Accidents: Living with High-Risk Technologies* (revised edn), Princeton University Press.

Petersen, K.L. (2012) 'Risk analysis: a field within security studies?', *European Journal of International Relations*, 18(4): 693–717.

Pettersen, K. (2016) 'Understanding uncertainty: thinking through in relation to high-risk technologies', in A. Burgess, A. Alemanno and J.O. Zinn (eds) *Routledge Handbook of Risk Studies*, Routledge, pp 39–48.

Power, M. (2007) *Organized Uncertainty: Designing a World of Risk Management*, Oxford University Press.

Prakash, A. and Potoski, M. (2010) 'The International Organization for Standardization as a global governor: a club theory perspective', in D. Avant, M. Finnemore and S.K. Sell (eds) *Who Governs the Globe?*, Cambridge University Press, pp 72–101.

Radu, R. (2013) 'Power technology and powerful technologies: global governmentality and security in the cyberspace', in J.-F. Kremer and B. Müller (eds) *Cyberspace and International Relations: Theory, Prospects and Challenges*, Springer Berlin, pp 3–20.

Raissipour, D. (2023) 'Assessing the correlation between cyber risk and business risk', *Forbes*, 7 June, Available from: https://www.forbes.com/sites/forbestechcouncil/2023/06/07/assessing-the-correlation-between-cyber-risk-and-business-risk/

Rasmussen, M.V. (2006) *The Risk Society at War: Terror, Technology and Strategy in the Twenty-First Century*, Cambridge University Press.

Reith, G. (2004) 'Uncertain times: the notion of "risk" and the development of modernity', *Time and Society*, 13(2–3): 383–402.

Renn, O. (2008) *Risk Governance: Coping with Uncertainty in a Complex World*, Routledge.

Renn, O., Lucas K., Haas A. and Jaeger, C. (2019) 'Things are different now: the challenge of global systemic risks', *Journal of Risk Research*, 22(4): 401–15.

Rodban, M. (2025) 'Political risk and the geopolitics of cybersecurity', in C.E. Sottilotta, J. Campisi, J. Leitner and H. Meissner (eds) *The Routledge Handbook of Political Risk*, Routledge, pp 156–71.

Romaniuk, S.N. and Manjikian, M. (eds) (2021) *Routledge Companion to Global Cyber-Security Strategy*, Routledge.

Romanosky, S., Ablon, L., Kuehn, A. and Jones, T. (2019) 'Content analysis of cyber insurance policies: how do carriers price cyber risk?', *Journal of Cybersecurity*, 5(1): tyz002.

Rosa, H. (2020) *The Uncontrollability of the World*, Polity.

Ross, R., Pillitteri, V., Graubart, R., Bodeau, D. and McQuaid, R. (2021) 'Developing cyber-resilient systems: a systems security engineering approach', National Institute of Standards and Technology Special Publication 800-160, vol 2, revision 1, Available from: https://nvlpubs.nist.gov/nistpubs/SpecialPublications/NIST.SP.800-160v2r1.pdf

Ruan, K. (2019) *Digital Asset Valuation and Cyber Risk Measurement: Principles of Cybernomics*, Academic Press.

Saalman, L. Su, F. and Dovgal, L.S. (2024) 'Cyber risk reduction in China, Russia, the United States and the European Union', Stockholm International Peace Research Institute, Available from: https://www.sipri.org/sites/default/files/2024-06/cyber_risk_reduction.pdf

Samonas, S. and Coss, D. (2014) 'The CIA strikes back: redefining confidentiality, integrity and availability in security', *Journal of Information System Security*, 10(3): 21–45.

Sandman, P. (1989) 'Hazard versus outrage in the public perception of risk', in V.T. Covello, D.B. McCallum and M.T. Pavlova (eds) *Effective Risk Communication: The Role and Responsibility of Government and Nongovernment Organizations*, Springer, pp 45–9.

Sangari, S., Dallal, E. and Whitman, M. (2022) 'Modelling under-reporting in cyber incidents', *Risks*, 10(11): article 200.

Santaniello, M. (2025) 'Attributes of digital sovereignty: a conceptual framework', *Geopolitics*, Available from: https://doi.org/10.1080/14650045.2025.2521548

Schmoldt, J. (2024) 'Cyber proxies: covert state–non-state interactions in cyberwarfare', in T. Stevens and J. Devanny (eds) *Research Handbook on Cyberwarfare*, Edward Elgar, pp 131–47.

Schneider, J. (2019) 'Persistent engagement: foundation, evolution and evaluation of a strategy', Lawfare, 10 May, Available from: https://www.lawfaremedia.org/article/persistent-engagement-foundation-evolution-and-evaluation-strategy

Schueler, C. (2024) 'Navigating the world of cyber insurance: is it worth it?', *Forbes*, 11 July, Available from: https://www.forbes.com/councils/forbestechcouncil/2024/07/11/navigating-the-world-of-cyber-insurance-is-it-worth-it/

Scroxton, A. (2025) 'UK's Cyber Monitoring Centre begins incident classification work', *Computer Weekly*, 6 February, Available from: https://www.computerweekly.com/news/366618805/UKs-Cyber-Monitoring-Centre-begins-incident-classification-work

Shaw, M. (2005) *The New Western Way of War: Risk-Transfer War and Its Crisis in Iraq*, Polity.

Shires, J. (2019a) 'Family resemblance or family argument? Three perspectives on cybersecurity and their interactions', *St. Anthony's International Review*, 15(1): 18–36.

Shires, J. (2019b) 'Hack-and-leak operations: intrusion and influence in the Gulf', *Journal of Cyber Policy*, 4(2): 235–56.

Shires, J. (2020) 'Ambiguity and appropriation: cybersecurity and cybercrime in Egypt and the Gulf', in D. Broeders and B. Van Den Berg (eds) *Governing Cyberspace: Behaviour, Power, and Diplomacy*, Rowman and Littlefield, pp 205–26.

Simpson, A. (2024) 'Into the unknown: the need to reframe risk analysis', *Journal of Cybersecurity*, 10(1): tyae022.

Skeoch, H.R.K. and Ioannadis, C. (2024) 'The barriers to sustainable risk transfer in the cyber-insurance market', *Journal of Cybersecurity*, 10(1): tyae003.

Slayton, R. and Muller, L.P. (2025) 'Coordinating uncertainty in the political economy of cyber threat intelligence', *Social Studies of Science*, 55(4): 481–511.

Sloan, R.H. and Warner, R. (2020) *Why Don't We Defend Better? Data Breaches, Risk Management, and Public Policy*, CRC Press.

Slovic, P., Finucane, M.L., Peters, E. and MacGregor, D.G. (2004) 'Risk as analysis and risk as feelings: some thoughts about affect, reason, risk, and rationality', *Risk Analysis*, 24(2): 311–22.

Slupska, J. (2021) 'War, health and ecosystem: generative metaphors in cybersecurity governance', *Philosophy and Technology*, 34(3): 463–82.

Smeets, M. (2018) 'The strategic promise of offensive cyber operations', *Strategic Studies Quarterly*, 12(3): 90–113.

Smeets, M. (2020) 'US cyber strategy of persistent engagement and defend forward: implications for the alliance and intelligence collection', *Intelligence and National Security*, 35(3): 444–53.

Smith, I. (2022) 'Cyber attacks set to become "uninsurable", says Zurich chief', *Financial Times*, 26 December, Available from: https://www.ft.com/content/63ea94fa-c6fc-449f-b2b8-ea29cc83637d

Spencer, N. and Pizio, D. (2024) 'The de-perimeterisation of information security: the Jericho Forum, zero trust, and narrativity', *Social Studies of Science*, 54(5): 655–77.

Stanham, L. (2023) 'Machine learning (ML) and cybersecurity', CrowdStrike, 2 November, Available from: https://www.crowdstrike.com/en-us/cybersecurity-101/artificial-intelligence/machine-learning/

Stanton, B., Theofanos, M.F., Prettyman, S.S. and Furman S. (2016) 'Security fatigue', *IT Professional*, 18(5): 26–32.

Star, S.L. (1999) 'The ethnography of infrastructure', *American Behavioral Scientist*, 43(3): 377–91.

Steed, D. (2019) *The Politics and Technology of Cyberspace*, Routledge.

Steger, M.B. (2008) *The Rise of the Global Imaginary: Political Ideologies from the French Revolution to the Global War on Terror*, Oxford University Press.

Stetter, S. (2021) 'Global governmentality and Foucault's toolbox: reflections on international politics as a social system and field of power relations', in J. Busse (ed) *The Globality of Governmentality: Governing an Entangled World*, Routledge, pp 29–49.

Stevens, C. (2021) 'Negotiating a symbolic and organisational thing of boundaries', PhD thesis, University of Bristol, Available from: https://research-information.bris.ac.uk/en/studentTheses/bounding-us-cybersecurity

Stevens, T. (2016) *Cyber Security and the Politics of Time*, Cambridge University Press.

Stevens, T. (2020) 'Knowledge in the grey zone: AI and cybersecurity', *Digital War*, 1(1–3): 164–70.

Stevens, T. (2021) 'United Kingdom: pragmatism and adaptability in the cyber realm', in S.N. Romaniuk and M. Manjikian (eds) *Routledge Companion to Global Cyber-Security Strategy*, Routledge, pp 191–200.

Stevens, T. (2023a) *What Is Cybersecurity For?*, Bristol University Press.

Stevens, T. (2023b) 'The map is not the territory: the epistemic uncertainty of cyber risk assessment', Security Lancaster Research Seminar, 1 November, Available from: https://kclpure.kcl.ac.uk/ws/portalfiles/portal/255147268/Stevens_The_Map_is_Not_the_Territory_Nov_2023.pdf

Stevens, T. and Kavanagh, C. (2021) 'Cyber power in international relations', in P. Cornish (ed) *The Oxford Handbook of Cyber Security*, Oxford University Press, pp 66–81.

Strupczewski, G. (2021) 'Defining cyber risk', *Safety Science*, 135: 105143.

Sukumar, A., Broeders, D. and M. Kello (2024) 'The pervasive informality of the international cybersecurity regime: geopolitics, non-state actors and diplomacy', *Contemporary Security Policy*, 45(1): 7–44.

Swiss Re (2024) 'Reality check on the future of the cyber insurance market', 18 November, Available from: https://www.swissre.com/risk-knowledge/advancing-societal-benefits-digitalisation/about-cyber-insurance-market.html

Tabansky, L., Cornish, P. and Lichterman, E. (2024) 'Making PROGRESS: a sectoral approach to cyber resilience and its application in sustainable development', *Journal of Cyber Policy*, 9(3): 423–40.

Tarun, R. (2023) 'When managing cybersecurity, operate like you've already been compromised', *Forbes*, 13 March, Available from: https://www.forbes.com/sites/forbestechcouncil/2023/03/13/when-managing-cybersecurity-operate-like-youve-already-been-compromised/?sh=32951b7a6b84

Tatar, U., Nussbaum, B., Gokce, Y. and Keskin, O.F. (2021) 'Digital force majeure: the Mondelez case, insurance, and the (un)certainty) of attribution in cyberattacks', *Business Horizon*, 64(6): 775–85.

Thomas, H. (2023) 'The UK will need more than words in this cyber war', *Financial Times*, 20 April, Available from: https://www.ft.com/content/a70d4af7-1006-409e-9e78-da0114c10a0a

Thornton, J. (2024) 'Exponential data growth: more data, less gains', Big Data Plumbing, 23 July, Available from: https://bigdataplumbing.substack.com/p/exponential-data-growth

Tiirmaa-Klaar, H. (2016) 'Building national cyber resilience and protecting critical information infrastructure', *Journal of Cyber Policy*, 1(1): 94–106.

Tzavara, V. and Vassiliadis, S. (2024) 'Tracing the evolution of cyber resilience: a historical and conceptual review', *International Journal of Information Security*, 12: 1695–719.

UK Cyber Security Council (nd) 'Cyber security career pathways', Available from: https://www.ukcybersecuritycouncil.org.uk/careers-and-learning/cyber-career-framework/

US Cybersecurity and Infrastructure Security Agency (nd) 'Systemic cyber risk reduction', Available from: https://www.cisa.gov/resources-tools/programs/systemic-cyber-risk-reduction

US Joint Chiefs of Staff (2018) 'Cyberspace operations', Joint Publication 3–12, Available from: https://irp.fas.org/doddir/dod/jp3_12.pdf

Van Loon, J. (2002) *Risk and Technological Culture: Towards a Sociology of Virulence*, Routledge.

Vaughan-Williams, N. and Stevens, D. (2016) 'Vernacular theories of everyday (in)security: the disruptive potential of non-elite knowledge', *Security Dialogue*, 47(1): 40–58.

Villegas, S. (2025) 'Ciaran Martin: nobody plays politics with cyber', *Holyrood*, 26 February, Available from: https://www.holyrood.com/news/view,ciaran-martin-nobody-plays-politics-with-cyber

Von Lucke, F., Wellmann, Z. and Diez, T. (2014) 'What's at stake in securitising climate change? Towards a differentiated approach', *Geopolitics*, 19(4): 857–84.

Vuorinen, J. and Uusitupa, V. (2022) 'The emergence of liminal cyberspace: challenges for the ontological work in cybersecurity', in P. Kommers and M. Macedo (eds) *Proceedings of the International Conferences on ICT, Society and Human Beings 2022, Web Based Communities and Social Media 2022 and e-Health 2022*, IADIS Press, pp 96–103.

Walker, R.B.J. (1988) *One World, Many Worlds: Struggles for a Just World Peace*, Lynne Rienner.

Walters, W. (2012) *Governmentality: Critical Encounters*, Routledge.

Weinberger, D. (2012) *Too Big to Know: Rethinking Knowledge Now that the Facts Aren't the Facts, Experts are Everywhere, and the Smartest Person in the Room is the Room*, Basic Books.

Weiss, M. and Jankauskas, V. (2019) 'Securing cyberspace: how states design governance arrangements', *Governance*, 32(2): 259–75.

Weiss, M. and Krieger, N. (2025) 'The political economy of cybersecurity: governments, firms and opportunity structures for business power', *Contemporary Security Policy*, 46(3): 403–28.

Welburn, J.W. and Strong, A.M. (2022) 'Systemic cyber risks and aggregate impacts', *Risk Analysis*, 42(8): 1606–22.

Welch, D.A. (2022) *Security: A Philosophical Investigation*, Cambridge University Press.

Whannel, K. (2025) 'UK must prepare for war scenario, government warns', BBC News, 24 June, Available from: https://www.bbc.co.uk/news/articles/cpqnlxr43zdo

White House (2023) 'National cybersecurity strategy', Available from: https://bidenwhitehouse.archives.gov/wp-content/uploads/2023/03/National-Cybersecurity-Strategy-2023.pdf

Whiting, A. (2020) *Constructing Cybersecurity: Power, Expertise and the Internet Security Industry*, Manchester University Press.

Whyte, C. (2023) 'Learning to trust Skynet: interfacing with artificial intelligence in cyberspace', *Contemporary Security Policy*, 44(2): 308–44.

Whyte, C., Thrall, A.T. and Mazanec, B.M. (eds) (2021) *Information Warfare in the Age of Cyber Conflict*, Routledge.

Willett, M. (2024) *Cyber Operations and Their Responsible Use*, International Institute for Strategic Studies.

Williams, M.J. (2008) '(In)security studies, reflexive modernization and the risk society', *Cooperation and Conflict*, 43(1): 57–79.

Wolff, J. (2022) *Cyberinsurance Policy: Rethinking Risk in an Age of Ransomware, Computer Fraud, Data Breaches, and Cyberattacks*, MIT Press.

Wolff, J. (2024) 'The role of insurers in shaping international cyber-norms about cyber-war', *Contemporary Security Policy*, 45(1): 141–70.

Woods, D.W. and Wolff, J. (2025) 'A history of cyber risk transfer', *Journal of Cybersecurity*, 11(1): tyae028.

Work, J.D. (2020) 'Evaluating commercial cyber intelligence activity', *International Journal of Intelligence and CounterIntelligence*, 33(2): 278–308.

World Economic Forum (WEF) (2025a) 'The global risks report 2025: 20th edition insight report', Available from: https://reports.weforum.org/docs/WEF_Global_Risks_Report_2025.pdf

World Economic Forum (WEF) (2025b) 'Global cybersecurity outlook 2025: insight report', Available from: https://reports.weforum.org/docs/WEF_Global_Cybersecurity_Outlook_2025.pdf

Yoffe, L., Matania, E. and Sommer, U. (2025) 'The rise of responsible behavior: Western commercial reports on Western cyber threat actors', *Contemporary Security Policy*, 46(3): 429–54.

Young, D., Lopez Jr, J., Rice, M., Ramsey, B. and McTasney, R. (2016) 'A framework for incorporating insurance in critical infrastructure cyber risk strategies', *International Journal of Critical Infrastructure Protection*, 14: 43–57.

Zeng, J., Stevens, T. and Chen, Y. (2017) 'China's solution to global cyber governance: unpacking the domestic discourse of "internet sovereignty"', *Politics and Policy*, 45(3): 432–64.

Index

D

Printed and bound by CPI Group (UK) Ltd, Croydon, CR0 4YY

15/09/2026

14962386-0003